# Motions of Celestial Bodies

Computer simulations

# Motions of Celestial Bodies

Computer simulations

**Eugene Butikov**

*Saint Petersburg State University, Russia*

**IOP** Publishing, Bristol, UK

ISBN 978-0-750-31100-7 (ebook)
ISBN 978-0-750-31101-4 (print)
ISBN 978-0-750-31109-0 (mobi)

DOI 10.1088/978-0-750-31100-7

Version: 20141001

IOP Expanding Physics
ISSN 2053-2563 (on-line)
ISSN 2054-7315 (print)

British Library Cataloguing-in-Publication Data: A catalogue record for this book is available from thc British Library.

Published by IOP Publishing, wholly owned by The Institute of Physics, London

IOP Publishing, Temple Circus, Temple Way, Bristol, BS1 6HG, UK

US Office: IOP Publishing, Inc., 190 North Independence Mall West, Suite 601, Philadelphia, PA 19106, USA

# Contents

# Preface

This textbook *Motions of Celestial Bodies: Computer Simulations*, together with the accompanying award-winning educational software package PLANETS AND SATELLITES, is intended to help students learn and understand the fundamental concepts and laws of physics as they apply to the fascinating world of the motions of natural and artificial celestial bodies. In this wonderful space laboratory, all phenomena are observed in their purest form, without the numerous complications that are inevitable in an ordinary Earth laboratory. It is the understanding of the foundations of classical and modern physics that form the primary aim of the book, while their application to real-world celestial mechanics is more illustrative and incidental.

This textbook relies heavily on PLANETS AND SATELLITES, which includes several highly interactive computer programs presenting a set of exciting computer-simulated experiments. The programs of the package provide students and their instructors with a powerful tool which enables them to investigate basic concepts and phenomena that are difficult to imagine and study in a conventional abstract manner.

This textbook along with PLANETS AND SATELLITES has been developed as an exploration-oriented complement to various physics courses. It can be helpful to a wide range of students, from those in introductory physics to those in advanced courses. With this textbook and the software, the students can learn the basic principles and concepts of classical dynamics and the application of these principles to the motions of various celestial bodies—stars, planets, comets, natural and artificial satellites, and manned and automatic space vehicles. The simulation programs make visible the beauty and aesthetics of the mathematics and the fundamental laws of physics in their application to the motions of celestial bodies.

PLANETS AND SATELLITES allows the students to construct and investigate a model of the Solar System or to create an imaginary planetary system on their own—complete with the star, planets, moons, comets, asteroids and satellites. Contemporary interactive media provides students with a powerful means to visualize the evolution of such a planetary system and to explore the orbital motions governed by the gravitational forces. The simulations bring to life many abstract concepts of classical dynamics.

Interactive work with PLANETS AND SATELLITES helps students understand phenomena better. Students can work at a pace they enjoy, varying the parameters of the simulated systems and repeating the most interesting experiments several times on their own. Experience based on students' own actions results in deeper understanding than merely receiving someone else's knowledge. Undoubtedly, for the great majority of humans, a visual experience is much more intensive and permanent than an aural or reading experience. With some of the suggested programs, students have an opportunity to perform interesting mini-research projects in physics and astronomy.

The computer simulations in PLANETS AND SATELLITES enable students to see clearly how systems that obey simple and precise physical laws behave, sometimes in unexpected and even irregular, chaotic ways. Although designed as a desk-top laboratory for individual interactive work, the software also provides the instructor with powerful demonstration tools to accompany lectures in mechanics and general physics.

The structure of the programs allows students to study the subject at different levels of difficulty, depending on the time available and on the mathematical complexity of the course. Part I (chapters 2–9) of the textbook contains a description of the simulation programs and their possibilities, and explains how the programs are operated. It suggests experiments that demonstrate typical examples of the behavior of the simulated systems. Part I is aimed at building physical intuition. Understanding the underlying concepts of physics is given precedence over using formulas in calculations. Students who are not going to study the subject thoroughly may restrict themselves to only this part of the textbook. In order to remain simple, the text focuses on concepts while keeping mathematics to the necessary minimum. However, to answer some of the suggested questions and to solve the more difficult problems, students may also need to read the following parts. Once they master the basics, the structure of the textbook and software allows for increasingly complex investigations of the subject matter.

Chapter 10 of part II presents a more detailed, though rather elementary, description of the fundamental concepts and the laws of physics underlying the simulated phenomena. Quantitative, mathematical formulation of the physical laws and their consequences provides a means to predict and explain the results of the simulation experiments and to calculate the parameters that must be entered in order to obtain the desired results.

Chapter 11 of part II is much more sophisticated and is intended for an in-depth study of the subject. This highly mathematical chapter delves into the serious theoretical background for the computer-aided study of celestial mechanics and space dynamics.

In the sections 'Questions and problems' an asterisk (*) after the problem number is used to grade the problems by difficulty. Simple problems and questions have no asterisk. One asterisk means that the problem requires thought. Two- and especially three-asterisk problems are challenging in some way. The best learning takes place when you struggle with a tough problem.

Eugene Butikov
St Petersburg State University, Russia

# Author biography

## **Eugene Butikov**

Eugene Butikov was born on 17 January 1940 in Yekaterinburg (formerly Sverdlovsk), Russia. He graduated from Saint Petersburg (formerly Leningrad) State University in 1962 (Department of Physics). All of his research and academic activity took place at Saint Petersburg State University and Saint Petersburg Research University of Fine Mechanics and Optics, where presently he is a full professor of general physics. He lectures on general physics, optics, the quantum theory of solids, and the theory of oscillations. He has written several textbooks and handbooks on physics which are used widely in Russia.

His research work is associated with solid state physics (quantum theory of electronic paramagnetic resonance, theory of Josephson effects in weak superconductivity), theory of nonlinear oscillations. Several new complicated and even counterintuitive modes of regular and chaotic behavior have been discovered recently in parametrically excited simple nonlinear systems by virtue of his efforts with the help of computer simulations. He has succeeded in finding clear physical explanations for some of these modes, and in a theoretical determination of their boundaries in the parameter space of the system.

During the last decades he has devoted much time and effort to developing (and using) interactive educational software for university-level physics students. His software is a kind of desk-top laboratory in which the computer is used as a universal research instrument for investigating mathematical models of physical systems rather than simply as a convenient means to search and display information. He has developed several packages of interactive simulation programs with detailed tutorials. These programs are published and distributed in Russia by the Physicon Company, and in the USA by the American Institute of Physics. With these packages, he has tried to change the traditional paradigm of teaching physics in Universities. Students do most of the work on their own, using detailed manuals with assignments, which prompt them to do particular activities to explore the use of the simulation programs. Their work on those assigned projects is a model of real scientific research.

**IOP** Publishing

Motions of Celestial Bodies
Computer simulations
**Eugene Butikov**

# Chapter 1

## Introduction: getting started

In the introduction, the simulation programs available in the package PLANETS AND SATELLITES are listed, with information on how to operate the programs and how to execute commands. Part I of the textbook contains an elementary introduction to celestial mechanics based on computer simulations of motions of natural and artificial celestial bodies. It suggests activities and experiments to be performed with the interactive programs of the package PLANETS AND SATELLITES in order to illustrate the basic concepts and phenomena without using much mathematics. It also provides a detailed description of the simulation programs and their possibilities. Part II provides a serious theoretical background for the simulated phenomena and is intended for in-depth study of the subject.

### 1.1 List of the simulation programs

- Tutorial and overview of the simulations
- Kepler's laws
  - Kepler's first law
  - Kepler's second law
  - Kepler's third law
- Hodograph of the velocity vector
- Missiles and satellites
- Active maneuvers in space orbits
- Precession of an equatorial orbit
- Two-body and many-body systems
  - Double star
  - Planet with a satellite
  - Double star with a planet
  - Planetary system

doi:10.1088/978-0-750-31100-7ch1

## 1.2 How to operate the simulation programs

The software package `PLANETS AND SATELLITES` is to be used with the Microsoft Windows operating system. Its graphical user interface makes all control functions very simple. The motion of the simulated physical systems—stars, planets, satellites—is displayed on the computer screen in windows. The positions and dimensions of these can be customized in the usual manner, i.e. by dragging with the mouse. Below is a brief overview of the basic control functions carried out with the help of the command buttons and menu items.

- The **menu bar** at the top of each window displays the commands used to operate the program. Also, right-clicking the mouse anywhere in the program window invokes a **pop-up menu** with several commands.
- The **command buttons** under the menu bar provide quick access to several commonly used commands. You click a command button once to carry out the action represented by that button.
- The **sliders** with the labels 'Speed Up' and 'Slow Down' allow you to vary the speed of animation for convenience of observation by changing the time scale in which the motion is simulated and displayed.

The **command buttons** common to most of the programs execute the following actions:

- **Start, Pause, Go**—starts the simulation, pauses the simulation, continues the interrupted simulation;
- **Restart**—restores the initial conditions and repeats the simulation from the beginning;
- **Erase**—(in many-body systems) clears the window (erases all traces) and continues the simulation.

The **menu items** common to all the programs carry out the following functions:

- **File**:
  - **Exit**—closes all windows and terminates the program;
  - **Print**—opens the panel for printer settings.
- **Input**—offers a choice of parameters for input or opens the panel for entering several parameters of the simulated system (and the initial conditions for the simulation).
- **Options**:
  - **Dark Sky**—displays the simulated system on a black background if the item is selected (recommended) and on a white background if not;
  - **Stars on the Sky**—shows the stars if the item is selected;
  - **Predicted Curve**—shows the theoretically calculated orbit (the osculating orbit for a perturbed motion) before the simulation (and when the command button 'Pause' is clicked) if the item is selected;
  - **Bright Head**—shows the moving body (satellite, planet) as a bright circle if the item is selected and as a colored circle if not;
  - **Traces**—draws the trajectories of the moving bodies (satellites, planets) if the item is selected;

  - **Thick Curves**—shows the traces using thick lines if the item is selected;
  - **Time Marks**—fixes the positions of the moving bodies on the screen after certain equal time intervals if the item is selected.
- **Examples**—offers a list of predefined examples to choose from or opens a panel with a set of examples provided.
- **Zoom**—opens another window with a larger ('Zoom In') or smaller ('Zoom Out') image of the system, or opens an additional window to display the system in a different reference frame.
- **Close** (in additional windows)—closes the window and returns to the main window.

Other controls specific to separate programs are described in the corresponding sections of the textbook. For more information on the controls and on how to use the simulation programs, search in 'How To...' under the menu 'Help' from within the programs. Context-sensitive physical explanations about the simulated systems can be summoned by pressing the F1 key or by clicking on the menu item 'Help on physics'.

A convenient way to launch appropriate simulations while simultaneously reading the textbook consists in opening the on-line tutorial (the 'Tutorial' button on the title screen of the program). There are buttons in each item of the tutorial that allow you to launch relevant examples illustrating the discussed phenomenon. If using these buttons, you do not need to find a suitable program or input any parameters.

## 1.3 Keplerian motions in celestial mechanics

The motions of the planets as they slowly travel across the sky have long fascinated mankind—at least since the beginning of recorded history. Observational investigation of the motion of celestial bodies is the subject of *astrometry,* a branch of *astronomy* concerned with the measurement of the positions of celestial bodies. Astronomy is older than physics. One can even argue that physics as a serious science began when very early astronomers brought order to the description of planetary motion by discovering an amazing simplicity in the motions of the planets. The later explanation of this simplicity may be regarded as the origin of modern physical science. The empirical discovery of laws of planetary motion by Johannes Kepler around 1600 and the dynamical explanation of these laws by Isaac Newton around 1685 form a remarkable chapter in the history of science. It was an astonishing leap in our understanding of nature achieved by giant intellects.

During the more than three centuries since Newton, very convincing observational support for the laws of classical *Newtonian mechanics* has been found in the motions of various celestial bodies—natural and artificial satellites of planets, asteroids and major planets of the Solar System, comets, multiple star systems and galaxies. In this wonderful space laboratory the fundamental physical phenomena are observed in their purest form, that is, without the complications of friction and air resistance that are inevitable in an ordinary Earth laboratory.

The role of experiment in celestial mechanics is played by observation. Only during the last few decades, with the construction and launching of satellites and spacecraft, has celestial mechanics become an experimental science. Artificial satellites orbiting the Earth and other planets, and space probes and scientific space stations exploring the outer regions of our Solar System are familiar facts of contemporary life.

The theoretical background of celestial mechanics and of its modern, rapidly developing branch—space dynamics, which is the study of the motion of artificial celestial bodies—is provided by *Newton's law of universal gravitation* and by *Newton's laws of motion*. These laws of motion are the basis of classical dynamics.

The differential equations of motion describing some of the physical systems simulated in the suggested programs have analytic solutions. Such solutions occur for the motion of a body under the gravitational pull of another body whose mass is much greater. The motion of a planet around the Sun or a satellite around the Earth provide examples of such a *single-body problem*.

Exact analytic solutions also exist for the motion of two celestial bodies of comparable masses attracted by their mutual gravitational force. The theoretical investigation of this motion is referred to as the *two-body problem*. Mathematically, the two-body problem may be reduced to the problem of a single body (of a certain so-called reduced mass) which moves in the gravitational field of the other body, considered to be fixed in space.

This gravitational field is an example of a central field known generally as a *Newtonian* or *Coulomb field*. Such fields have the property that the force on a point-body in the field is inversely proportional to the square of its distance from the center of the field. This kind of dependence on distance (inverse-square decrease) is characteristic of the electrostatic interaction between small charged bodies ('point charges'), described by Coulomb's law, as well as of the gravitational attraction of small massive bodies ('point masses'), described by Newton's law of gravitation.

The universal character of the inverse-square diminution with distance expressed by Newton's law of gravitation and by Coulomb's law of electrostatic interaction is probably related to the most general geometric properties of our three-dimensional physical space, in which the surface area of a sphere is proportional to the square of its radius. There must apparently exist a very convincing reason for the fact that the common inverse-square law is valid for the fundamental forces acting in physical systems so immensely different in size as the hydrogen atom and the gravitating masses of the Universe.

It is very important to this study that the inverse-square dependence of the gravitational force on distance is valid not only for point masses, that is, for bodies separated from one another by distances that are much greater than the linear dimensions of the bodies, but also for any bodies whose masses are distributed with spherical symmetry, regardless of the distance between them. The force of attraction in this case is also determined by the same simple formula of Newton's law of universal gravitation provided we substitute therein the geometric center-to-center distance between the bodies. In other words, in applying the gravitational interaction between spherically symmetric bodies, we can consider the mass of each body to be concentrated at its geometric center. This mathematical fact, for which Isaac Newton gave the first proof, is extremely important in the case of stars and planets, which are for most practical applications spherically symmetric.

Therefore the solution to *Kepler's problem*, that is, the problem of the motion of a mass point under the action of a central force with the inverse-square dependence on the distance, is applicable not only to the motion of a planet around the Sun, where the distance between the gravitating bodies is much greater than their dimensions, but also to the motion of satellites in low orbits around the planets. The gravitational field of a planet at any point outside the planet is practically the same as if all the mass were concentrated at the center of the planet. And certainly there is no need for the satellite or the spacecraft to be spherically symmetric, since its dimensions are in any case negligible compared to the distance between the satellite and the center of the planet. This means that we can assume the gravitational field of the planet to be homogeneous at the extent of the satellite, considering the gravitational force to be applied at its center of mass. In other words, we can consider an artificial satellite of any realizable size and shape to be a material point.

The striking mathematical simplicity of possible trajectories is a distinctive feature of Keplerian motion. Any possible motion in the Newtonian gravitational field occurs along one of the *conic sections* or *conics*—curves formed by the intersection of a circular cone by a plane. Depending on the inclination of the intersecting plane to the axis of the cone and the vertex angle of the cone, the curve formed by the section may be a circle (the intersecting plane is perpendicular to the axis of the cone), an ellipse, a parabola (the intersecting plane is parallel to a straight line lying in the lateral surface of the cone), or a hyperbola.

Periodic motions of planets and satellites occur along closed *elliptical* (in particular *circular*) orbits. One of the foci of an elliptical orbit is located at the center of the gravitational force (i.e. at the center of the Sun for planetary revolutions, or at the center of the Earth for an artificial satellite orbiting the Earth).

The *parabolic* trajectory corresponds to the limiting case of an extremely stretched elliptical orbit whose second focus is effectively at infinity. In this case, the body is virtually at rest when it is very far from the center of force ('at infinity'). That is, the speed of the body approaches zero as its distance increases without limit.

If a body approaches the center of force from infinity where it had a finite (nonzero) speed, the motion occurs along one of the two branches of a *hyperbola*. The body swings around the center of the gravitational force, and then moves away indefinitely along the other side of the hyperbola. The motion to infinity along a branch of a hyperbola also occurs if a body at a finite distance from the center of force has a large enough speed, being accelerated to that speed, say, by a rocket engine. The initial speed needed for this purpose must exceed some value called the *escape velocity*. In this case the speed approaches a nonzero value (called the *hyperbolic excess of velocity*) as the distance increases without limit.

## 1.4 Numerical and analytical methods

Exact analytic solutions to the differential equations of motion, like the solution of Kepler's problem for the motion in a Newtonian gravitational field, are remarkable for the simplicity of motions described by these solutions. Unfortunately, exact solutions are seldom encountered in physics. When there are perturbing interactions (gravitational forces produced by other planets, deviations of mass distributions from exact spherical symmetry, etc), the equations of motion become non-integrable. The marvel of the closed orbits found in Keplerian motion, as well as their wonderful simplicity, vanishes. The mathematical analysis of such perturbed motions is immensely complicated.

When the perturbing forces are small compared to the main gravitational force, one can use approximate analytic methods. Keplerian motion in this case can be assumed as the zeroth approximation to the actual motion. We can consider that the small perturbations cause relatively slow variations of the parameters characterizing the corresponding Keplerian orbit, and try to obtain analytic expressions for these slow variations. Such an ellipse with varying parameters, grazing the actual trajectory, is called the *osculating* orbit. If we imagine that the perturbations suddenly vanish, all parameters of the osculating ellipse remain constant during the subsequent motion and the body traces the ellipse which touches the actual trajectory at the given point. This unperturbed Keplerian motion, for at least some part of the osculating ellipse, is very close to the actual motion.

However, when it is inadmissible to regard the perturbations as small, as, for example, in the general case of the three-body problem, it is impossible to obtain even approximate analytic solutions. In such cases, to solve the differential equations of motion, we can rely only on numerical methods.

It is very easy to understand the general idea of the numerical methods for the calculation of motion for a system of interacting bodies. For some initial moment let the position and velocity of the body under consideration be given. Also given are the positions and velocities of all the celestial bodies whose forces of gravitation cause the acceleration of the body in question. On the basis of Newton's law of gravitation and Newton's law of motion (Newton's second law), we calculate the gravitational acceleration of the given body caused by each of the celestial bodies separately and thus the total acceleration as the vector sum of these individual accelerations. Knowing the magnitude and direction of the initial velocity of the body, we calculate the new velocity and position of the body after a short time interval (after the temporal 'step' of integration). Similarly, we calculate the new velocities and positions of all the other bodies of the system after this time interval.

For these new positions, we calculate the new acceleration of the given body caused by all the celestial bodies and also the new accelerations of all the celestial bodies. Then, using the same scheme, we calculate the new positions of the bodies and their velocities after the next short time interval, and so on. Thus we follow step by step the motion of the system of interacting bodies during any finite time interval.

The only approximation that we make in this calculation is the assumption that during each small time interval (during the step of calculation) the acceleration of the body is constant. Actually it is changing continuously. We can improve the precision of the calculation by decreasing the step of integration. To be sure, this improvement is achieved at the cost of a greater number of calculations.

We have described above the simplest algorithm of integration known as Euler's scheme. This method provides relatively low precision and generates accumulating errors. That is why in the simulation methods of the suggested programs in the `PLANETS AND SATELLITES` package, a somewhat more complex algorithm is used (the Runge–Kutta method of the fourth order). This algorithm is free of the above mentioned deficiencies in Euler's method.

In all the programs of this package the simulation of motions of planets and satellites is based on the numerical integration of the equations of motion. Numerical integration is used even in cases for which analytic solutions are available. Analytic solutions are used only for drawing the theoretically predicted trajectories (if the corresponding option is chosen by the user), and for the automatic determination of a suitable scale for the representation of the motion on the computer screen. In other words, Kepler's laws and the laws of conservation are not used in the simulation experiments that are performed with the help of the programs. This is not required since the algorithm of numerical integration is based only on Newton's law of motion and Newton's law of gravitation.

All this means that we can treat the coincidence of the motion simulated in the programs with the corresponding theoretical predictions as an *experimental verification* of Kepler's laws (by means of a numerical computational experiment) for the motion of a body in a central gravitational field, and for the motion of two interacting bodies. Or, vice versa, if Kepler's laws are considered as reliable experimental facts established, say, by numerous astronomical observations, these computer simulations can be treated as additional experimental evidence for Newton's laws of motion which are the basis of classical dynamics.

If a third body is added to a system of two interacting bodies, the problem generally becomes analytically unsolvable. In other words, there exist no general formulas that describe the motion of the bodies and that allow us to calculate their positions from arbitrary initial conditions. Moreover, we cannot make conclusions about general properties of their motions. Evidently, the reason for the lack of analytical solutions for the three-body (and many-body) problem is related not only to the mathematical difficulties of the problem, but equally to the extraordinary complexity of the motions themselves. Some examples of such complex motions governed by mathematically a rather simple physical law (Newton's law of universal gravitation) can be found in several simulations executed with the help of the programs of the `PLANETS AND SATELLITES` package.

However, we encounter no fundamental difficulties when solving the three-body problem numerically. Solving the three-body problem numerically is just as easy as solving the two-body problem. Only the number of calculations for a given time interval increases. Nevertheless, it should be kept in mind that these universal numerical methods are by no means all-powerful. For the calculation of the velocity and position of a body after a long time interval, we need to know the initial state of the system with great accuracy. Moreover, these numerical methods give us no means of investigating the most general long-time properties of the motions of celestial bodies.

# Part I

## Review of the simulations

# Chapter 2

# Kepler's laws

Three principal laws of motion under the central gravitational force inversely proportional to the square of the distance from the center of force (Kepler's laws) are illustrated in the first three simulation programs of the package PLANETS AND SATELLITES. The fourth program of this cycle ('Hodograph of the velocity vector') is devoted to an interesting property of Keplerian motion concerning the shape of its trajectory in velocity space.

## 2.1 Kepler's first law

According to Kepler's first law, the orbits of planets are ellipses with the Sun at one focus of the ellipse. Although the law originally applied only to planets, it is also valid for any body whose motion is finite and is governed by a central force of attraction whose magnitude decreases in proportion to the square of the distance of the body from the center of force. In particular, the gravitational motion of natural and artificial satellites around the Earth and other planets also obeys Kepler's first law: in the reference frame of the planet the orbits of these satellites are ellipses or circles.

The simulation program called 'Kepler's First Law' uses one of the main geometric properties of an ellipse to demonstrate experimentally the elliptic form of the orbit. This property is that for any point on an ellipse the sum of the distances of the point to two fixed points (called the foci of the ellipse) is a constant equal to the longest diameter of the ellipse. This property is often used as a definition of the ellipse.

When the 'Go' button is clicked, the motion of a planet (or satellite) and its orbit, calculated numerically under the assumption that the force decreases as the square of the distance, is continuously displayed in the left window on the computer screen. The two straight lines that join the planet with the center of force and the second focus are simultaneously drawn after equal intervals of time as the planet moves in

doi:10.1088/978-0-750-31100-7ch2  

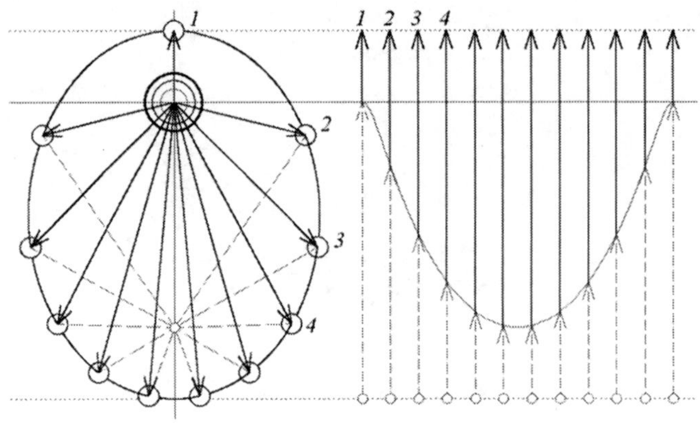


**Figure 2.1.** The Keplerian orbit (left) of a satellite and the geometric addition of its distances from the two foci (right). Corresponding numbers on the left and right diagrams refer to the same instants of time.

its orbit[1]. These straight lines (segments) are shown in different colors on the computer screen.

In the right window of the screen, these same segments are simultaneously redrawn, but here they are aligned along the vertical and placed one after the other so that we can easily add their lengths visually. We see that for all points of the closed orbit the sum of the lengths of these segments has the same value. This value is equal to the major axis of the ellipse. Figure 2.1 shows an example of the final image on the computer screen.

Since the program calculates the orbit only on the basis of Newton's laws of motion and the law of universal gravitation, we can treat the result of the simulation not only as a simple illustration, but rather as an experimental verification of Kepler's first law (in a computational experiment).

The experiment can be repeated with different magnitudes of the initial velocity, whose direction in this program is always transverse, that is, perpendicular to the initial radius vector. (The radius vector is the line directed from the center of force to the orbiting body.) The value of the initial velocity must be specified in units of the circular velocity for the given initial point. That is, if we enter 1, the body moves in a circle whose center is at the center of force. If a value greater than unity is entered, the initial point is the perigee (or the perihelion if the motion of a planet around the Sun is simulated in the experiment). That is, this initial point of the orbit is the nearest one to the center of force. The center of force is located at the near focus of the ellipse.

On the other hand, if the initial velocity is smaller than the circular velocity, the initial point is the apogee of the orbit (or aphelion if the motion of a planet is simulated); the center of force is located at the remote focus of the ellipse.

---

[1] The program determines the position of the second focus beforehand with the help of formulas based on the conservation laws and on the assumption that the orbit is an ellipse. However, the geometric proof displayed by the program is quite convincing independent of how the point has been found because the only important feature is that such a point exists.

On the right side of the screen, the equally spaced vertical segments are equal to the distance of the orbiting body from the center of force. These segments correspond to the instants that are separated by equal time intervals. Therefore, the curve joining the ends of these segments (see figure 2.1) can be treated as a graph of the time dependence of the distance from the center of force for the body moving along the orbit. (The ordinate axis of this graph is directed downward.)

Watching the simulation, we can also judge how the velocity of the body changes during the motion along the orbit. This motion is displayed on the screen in a time scale that is constant during the simulation. For convenience of observation, this time scale can be varied between certain limits by dragging the slide on the corresponding scroll-bar. Moreover, positions of the body along the trajectory are fixed on the screen after equal time intervals, and therefore the final static picture obtained in the experiment (see figure 2.1) also shows how the velocity changes along the orbit. We note that the marked positions of the body are sparse near the perigee (perihelion), where the motion is fastest, and dense near the apogee (aphelion), where the motion is slowest.

Especially noticeable is the variation in the angular velocity of rotation of the radius vector as the body moves along the orbit. There is a considerable difference in the size of the angles between adjacent radius vectors near the apogee and the size of the corresponding angles near the perigee (see figure 2.1). This non-uniformity of motion along an elliptical orbit is described quantitatively by Kepler's second law, illustrated in the simulation program discussed in the next section.

There is an option in the program which allows the theoretically predicted orbit to be drawn before the simulation is begun. The curve is plotted by the program on the assumption that it is exactly an ellipse defined by equation (10.14) (see section 10.4). The orbital parameter $p$ and eccentricity $e$ of the ellipse are calculated on the basis of the initial conditions that have been entered. Congruence of the actual trajectory (obtained in the simulation experiment) with this theoretical curve can be treated as one more piece of evidence that the Keplerian orbit is an ellipse.

A closed elliptical (or circular) orbit corresponds to the case of negative total energy of the body (provided the gravitational potential energy is chosen to be zero at an infinite distance from the center of force). To simulate such a motion, a value of the initial velocity should be entered that does not exceed the escape velocity. However, the simulation program can also handle the cases of zero or positive total energy, for which the trajectory is respectively a parabola or a hyperbola. These cases are examples of a generalization of Kepler's first law which states that motion in a Newtonian central gravitational field always occurs along a conic section.

To consider the cases for which the total energy is zero or positive, we chose 'Parabolic motion' or 'Hyperbolic motion' respectively in the corresponding drop-down list. In the latter case we can enter a value of the initial velocity which exceeds the escape velocity. The corresponding trajectory is a hyperbola. To prove this, the program again draws straight lines from the foci to the points on the trajectory. Simultaneously in the right window these rectilinear segments are aligned next to one another so that we can visually subtract the length of one from the other. It is clear that the difference is the same for all points of the trajectory.

This demonstration proves that the curve is a hyperbola . The difference in lengths of the segments equals the distance between the vertices of the two branches of the hyperbola.

Since the direction of the velocity vector is transverse (perpendicular to the radius vector) at the initial position, this point is the vertex of the hyperbola along which the body recedes to infinity. Therefore, only half of this branch of the hyperbola is generated. In order to obtain the other half, the program also simulates the motion of the body from infinity toward the vertex of the hyperbola. This simulation is achieved in the following way. When the body recedes far enough from the initial position (beyond the region displayed on the screen), the program reverses the sign of one of its coordinates (namely, the sign of the coordinate measured in the direction of the initial velocity), and simultaneously reverses the other component of the velocity. Hence, subsequent motion of the body observed on the computer screen occurs toward the initial position along the other half of the same hyperbola.

If the option 'Predicted curve' is chosen, the program first draws the theoretically predicted hyperbolic trajectory together with the other branch of the hyperbola and the corresponding asymptotes. Parameters of the hyperbola are calculated in the program on the basis of the conservation laws for the value of the initial velocity that has been entered. The subsequent simulation of the motion does not use the results of this preliminary calculation—it is based entirely on a numerical integration of Newton's laws of motion. Therefore, coincidence of the actual simulated trajectory with the predicted curve can be treated as an additional experimental verification of the generalized Kepler's first law for a hyperbolic motion.

As the body moves away from the center of force, its trajectory gradually approaches the asymptote of the hyperbola. Eventually the motion is uniform rectilinear motion along the asymptote. The constant speed characteristic of this motion is called the hyperbolic excess of velocity. It equals the square root of the difference between the squares of the initial velocity and the escape velocity. The circles in the left window of the screen that fix spatial positions of the body after equal time intervals become equidistant. In the right window, the graph of the time dependence of the distance from the center of force asymptotically approaches a straight line.

A similar experiment can be performed for the parabolic motion, if the corresponding item is chosen from the drop-down list. In this case it is not necessary to enter the initial velocity since for the parabolic motion the initial velocity has a unique value, namely the value of the escape velocity characteristic of the initial position. The program sets this value automatically when the item 'Parabolic motion' is chosen. At the initial position this velocity is in the transverse direction. Consequently, the initial position is at the vertex of the parabola. Only half of the parabola is generated as the body recedes to infinity. To generate the motion along the other half of the parabola, the program makes the same changes in the position and velocity of the body as for the case of hyperbolic motion, when the body recedes beyond the area displayed on the screen (see above). The theoretically predicted parabolic curve is also shown before the simulation when the corresponding option is chosen.

Theoretical derivations of generalized Kepler's first law can be found in section 11.2, and section 11.5.

## Questions and problems

1. **Major axis of the orbit.** Let the initial velocity of a satellite at some height over the Earth's surface be directed horizontally. How is the major axis of its elliptical trajectory oriented with respect to the vertical line passing through the initial position? Does this orientation of the major axis depend on the magnitude of the initial velocity?
2. **Perigee and apogee of the orbit.** Where are the perigee and apogee of the elliptical orbit of a satellite located relative to the initial point if the velocity at this point is directed horizontally? Do the positions of these points depend on the magnitude of the initial velocity? Consider the cases in which the initial velocity is greater or smaller than the circular velocity for the given initial point.
3. **An ellipse or an oval.** Why can you conclude from the image displayed on the screen that the closed trajectory of a planet (or a satellite) is precisely an ellipse, and not an oval or some other closed curve?
4. **Foci of the orbit and the center of force.** One of the foci of the elliptical orbit coincides with the center of force. In which case is this focus of the ellipse nearest to the initial point of the orbit, and in which case is it the remote focus?
5. **Variation of the speed along the orbit.** At which point of an elliptical orbit is the speed of the planet (or satellite) greatest, and at which point is the speed smallest? Observe the variation in speed of the planet as it travels in its orbit and describe the characteristics of this variation.
6. **The displayed graph.** What is the physical sense of the curve obtained in the right window of the screen? This curve separates the segments whose lengths equal the distances of the planet (or satellite) from the foci of the elliptical orbit.
7. **The eccentricity of the elliptical orbit.** How does the eccentricity of the elliptical orbit resulting from a horizontal initial velocity depend on the magnitude of the initial velocity? How does the eccentricity of the elliptical orbit change as we increase the initial speed from zero to that producing a circular orbit and then on to the escape velocity?

   (**) For some value $v_0$ of the transverse initial velocity greater than the circular velocity ($v_0 > v_c$), we get an elliptical orbit whose nearest focus is located at the center of force. What should the other value of the initial velocity (smaller than the circular one) be in order that the resulting orbit be an ellipse homothetic to the first orbit?
8. **Axes of symmetry of a hyperbolic trajectory.** For the hyperbolic motion of a body arising from a sufficiently large initial horizontal velocity, how are the axes of symmetry of the hyperbola oriented with respect to the vertical line passing through the initial point? Does this orientation depend on the magnitude of the initial velocity? Does the angle between the asymptotes depend on the initial velocity?
9. (*) **The hyperbolic excess of velocity.** Let the initial velocity $v_0$ of a satellite be greater than the escape velocity $v_{esc}$ for the initial point. Prove that as the

satellite recedes to infinity, its speed tends to a constant value. Show that this value $v_\infty$ does not depend on the direction of the initial velocity and equals $\sqrt{v_0^2 - v_{esc}^2}$.

10. (**) **Angle between the asymptotes.** How does the angle between the asymptotes of a hyperbolic trajectory depend on the speed of the orbiting body at the vertex of the hyperbola (the point nearest to the center of force)? Assume that this speed is expressed in units of the circular velocity for this point.

    What should be the magnitude of the transverse initial velocity in order that the direction of motion changed through an angle $\theta$ as the body recedes from the initial point to infinity?

## 2.2 Kepler's second law

Kepler's second law, also known as the *law of equal areas*, governs the variations of the speed with which a planet (or a satellite) travels in its orbit. Kepler discovered this law in the early seventeenth century on the basis of a very careful examination of astronomical data collected by Tycho Brahe during his many years of observing the motion of the planets.

According to Kepler's second law, the radius vector directed from the Sun to the planet sweeps out equal areas in equal time intervals. In other words, the *areal* or *sectorial velocity* is constant during the orbital motion of a planet.

For a circular orbit this law means that a planet moves at a constant linear (and angular) speed. However, the motion of a planet along an elliptical orbit is non-uniform. The motion is fastest at perihelion and slowest at aphelion. Although Kepler's second law (as well as the first and third laws) was discovered through observations of planetary motion, it is also valid for satellites orbiting a planet.

Newton was the first to prove that for any curvilinear motion which satisfies the law of equal areas, the force accelerating the body in a curved path is always directed toward the same point. Conversely, if the force is directed toward the same center, the motion of the body obeys Kepler's second law. Thus, Kepler's second law is a consequence of the central character of forces acting between the Sun and the planets.

Hence the law of equal areas is valid not only for closed circular and elliptical orbits, but also for infinite parabolic and hyperbolic motions. Moreover, it is valid not only for any Keplerian motion under a central gravitational force which is inversely proportional to the square of the distance, but also for any motion in a central field with an arbitrary dependence of force on the distance.

Kepler's second law can be regarded as a consequence of the law of conservation of the angular momentum, which holds for any motion in an arbitrary central force field. An analytic derivation of this conservation law and of the law of equal areas (Kepler's second law) is found in section 11.1 of this textbook.

The simulation program allows us to verify Kepler's second law quantitatively by means of a computational experiment. The program calculates numerically the motion of a planet (or a satellite) under a central gravitational force that decreases with the square of the distance from the center of force. At the same time the radius

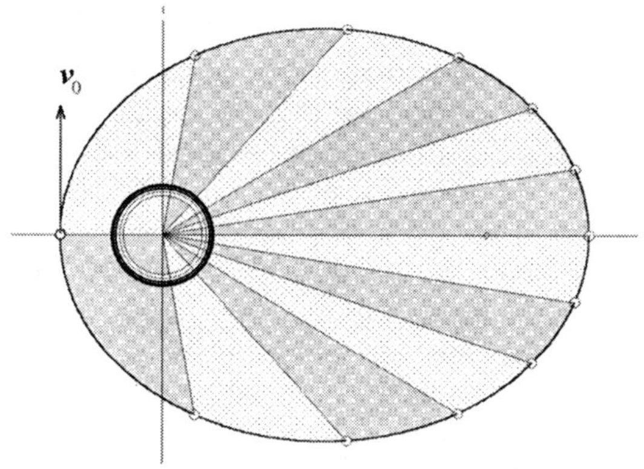


**Figure 2.2.** The areas swept out by the radius vector in equal time intervals for an elliptic Keplerian motion.

vector from the force center to the momentary position of the planet (or satellite) is drawn on the screen. After equal time intervals, the color of the radius vector changes. Hence adjacent sectors swept out by the radius vector in equal time intervals are painted in different colors on the screen. Since all the sectors correspond to equal time intervals, their areas, according to Kepler's second law, must be equal. Figure 2.2 shows the screen image when the orbit is completed.

It is difficult to judge visually whether the areas of sectors of different shapes swept by the radius vector in equal times are actually equal. In order for us to come to a quantitative conclusion, the program also calculates the areas of these sectors. In this calculation the elementary segments swept out by the radius vector during a step of integration are regarded as narrow triangles. The areas of these triangles (expressed in arbitrary units) are added until the sector is completely covered. The current (changing) value of the area is displayed on the screen. After a segment is swept out, the current area of the next-in-turn segment is displayed in the nearby position on the screen. The coincidence of the calculated areas for all filled segments testifies to the fulfilment of the law of equal areas.

We emphasize that the program does not 'know' anything about Kepler's laws. That is, it uses no information about the laws while working. The program only performs a numerical integration of the equations of motion (given by Newton's second law) under the central gravitational force. Therefore we can consider the equality of areas of segments obtained in this simulation as an experimental verification of Kepler's second law.

To demonstrate Kepler's second law for different orbits, we can vary in the program the magnitude of the initial velocity of the satellite. This velocity in the simulation program is transverse, that is, it is always directed horizontally (perpendicularly to the radius vector in the initial position). Consequently, the major axis of the elliptical orbit passes through the initial position and the center of the Earth.

The magnitude of the initial velocity which we enter in the program must be expressed in units of the circular velocity for the given initial point. The values exceeding 1 unit, that is, greater than the circular velocity (but smaller than the escape velocity $\sqrt{2} \approx 1.41$), produce orbits with the perigee at the initial position. If the initial velocity is set to be less than 1, the initial position is the apogee of the resulting orbit. It is assumed in this case that the initial position is sufficiently high above Earth that the orbit does not intersect the Earth's surface. The perigee of such an orbit is located over the opposite side of the Earth. The program chooses the scale automatically in order to fit the orbit into the window.

The program also displays the theoretically predicted orbit beforehand if the corresponding item from the menu 'Options' is selected.

To prove Kepler's second law for infinite trajectories, an initial velocity that equals or exceeds the escape velocity is to be entered. In these cases the initial position is the vertex of the parabolic or hyperbolic trajectory and is the point on the trajectory closest to the Earth. The speed of the satellite is greatest here.

If an initial velocity very close to the escape velocity is entered, the program converts it into the escape velocity. In this case the trajectory is a parabola.

When the satellite leaves the window, the program simulates the motion along the other half of the trajectory, as if the satellite were approaching the Earth from infinity. This is done in the same way as in the preceding program concerning the generalization of Kepler's first law for open orbits.

For hyperbolic trajectories, the program draws the theoretically predicted hyperbola and its asymptotes before plotting the trajectory through numerical calculation on the basis of the laws of motion (if the appropriate item in the menu 'Options' is selected). At large enough distances from the Earth, the satellite moves almost rectilinearly along the asymptote and with almost constant velocity. This velocity equals the square root of the difference between the squares of the initial velocity and the escape velocity. It is called the hyperbolic excess of velocity.

## Questions and problems

1. **Kepler's second law for different orbits.** What is the physical sense of Kepler's second law? How can we use it to describe the variation of the planet's speed (or of the satellite's speed) along the orbit? What does this law imply if we apply it to a circular orbit? To a closed elliptical orbit? To an infinite hyperbolic orbit?
2. (*) **Velocities at the perigee and apogee.** For a satellite orbiting the Earth in an elliptical orbit, what is the ratio of velocities at perigee and apogee? How does this ratio depend on the distances of these points from the center of the Earth? What is the ratio of the angular velocities of rotation of the radius vector for these points?
3. (**) **Lapse of time for half an orbit.** A satellite revolves around the Earth in an elliptical orbit whose eccentricity is 0.5. Consider the half of the orbit that is closest to the Earth. What is the ratio of the time taken for the satellite to cover this part of the orbit to the total period of revolution?
4. **Second law and the conservation of angular momentum.** How is Kepler's second law related to the law of the conservation of angular momentum? In order that Kepler's second law should hold, is it necessary that the force between a planet and the Sun be inversely proportional to the square of the distance separating them?

## 2.3 Kepler's third law

Kepler's third law was originally applied to the planets of the Solar System. It states that the squares of the periods of revolution of the planets around the Sun are proportional to the cubes of the major axes of their orbits. Like the first and second laws, this kinematic relationship between the periods of revolution and the linear dimensions of the orbits was discovered by Kepler from his careful analysis of numerous astronomical observations of the planets by Tycho Brahe. A derivation of Kepler's laws on the basis of the inverse-square law of universal gravitation was given by Isaac Newton about a half-century later.

Therefore, Kepler's third law is valid not only for the planets orbiting the Sun, but also for any group of satellites orbiting a common central body under the influence of a gravitational force of attraction: the squares of their periods of revolution are proportional to the cubes of the major axes of their orbits. An analytic proof is given for circular orbits in section 10.3, and for elliptical orbits in section 11.3.

The simulation provides several ways in which Kepler's third law may be experimentally verified. The simplest experiment involves only two satellites. Their motion is begun from the same initial position, where their velocities are transverse (perpendicular to the radius vector) (figure 2.3). One of their orbits is always circular and serves as a standard by which the period of revolution for the other, traveling in an arbitrary ellipse, is to be measured. The only quantity to be entered is related to this second orbit. The menu item 'Input' allows one of three options: the initial velocity $v_0$, the semimajor axis $a$, or the period of revolution $T$. Whichever of these parameters is chosen, it should be expressed in units that are characteristic of the standard circular orbit ($v_c$, $r_0$, or $T_0$, respectively).

The orbits graze one another at the initial position *1* (see figure 2.3). The program simultaneously simulates the motion of both satellites. The readings of the timer indicate the number of revolutions performed from the initial moment to the current moment for the satellite moving along the standard circular orbit. In other words, the timer shows the lapse of time in units of the period of revolution along the circular orbit.

The simulation of the motion halts automatically at the moment when the elliptical orbit is completed. Thus, the readings of the timer make it easy to compare the experimental values of the periods of revolution for the two orbits. For example,

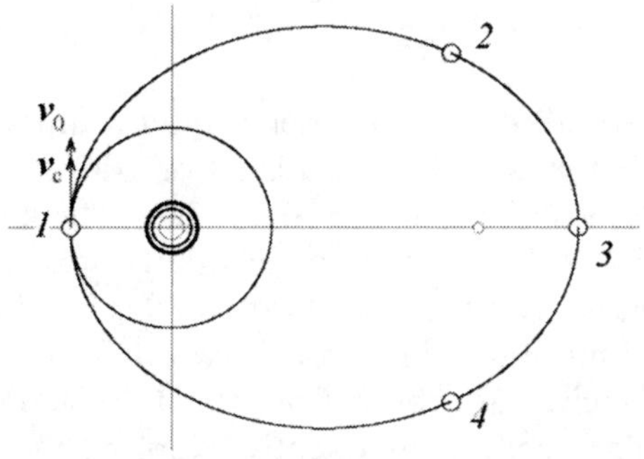


**Figure 2.3.** An elliptical orbit and a circular orbit with a common initial position.

if the value 4 is entered for the semimajor axis of the elliptical orbit (four times the radius of the circular orbit), the period of revolution for the elliptical orbit, according to Kepler's third law, equals eight periods of revolution of the circular orbit. In other words, during the revolution of the satellite around the ellipse, the other satellite in the circular orbit performs exactly eight revolutions. Both satellites return to the initial position simultaneously, thus reproducing the initial configuration after every eight revolutions of the satellite in the circular orbit.

We emphasize again that the simulation of motion in the program is based only on Newton's laws, not on Kepler's laws. The simulation may therefore be considered a test of the corresponding theoretical concepts.

A convenient way to display the results of the simulation experiment may be chosen from the menu item 'Options'. The option 'Predicted curve' displays the theoretical orbits (whose calculation is based on Kepler's laws) before the simulation of motion is performed by the program on the basis of Newton's laws. The simulation is displayed at a fixed time scale in order that the variations in speed can be easily observed. The rate of animation can be adjusted within certain limits by dragging the slide on the appropriate scroll-bar. The option 'Time marks' forces the program to fix positions of the satellite in the elliptical orbit each time the satellite in the circular orbit completes a revolution. Using these time marks, we can judge how the velocity of the satellite changes along the orbit even from the static picture that remains on the screen after the simulation is completed. The meaning of other options is clear from their names.

The menu item 'Orbits' suggests more sophisticated experiments related to Kepler's third law. The option 'Common initial point' allows the exploration of a family of elliptical orbits starting from the same point. The orbits correspond to satellites that are given transverse initial velocities of different magnitudes. A special panel can be opened for entering parameters. It is possible to enter either the initial velocities, the major axes, or the periods of revolution (depending on the option chosen) for several orbits at once. The quantities entered must be expressed in units corresponding to the standard circular orbit. We can then observe the simulation of motion of the satellites and measure their periods either in a group of sequential launches (for which each satellite in an elliptical orbit is viewed simultaneously with the satellite in the standard circular orbit), or in a group of simultaneous launches.

The first possibility is convenient for the precise experimental measurement of the period of revolution: when the option 'Sequential launches' is chosen, the simulation halts when the cycle of motion along an elliptical orbit is completed, and the timer is easily read.

The second possibility allows us to observe the evolution of the satellites' configurations continuously for as long as we wish. For example, we can display several orbits with multiple periods of revolution (figure 2.4). The initial velocity for orbit *2* (or the semimajor axis of the orbit) is deliberately chosen so that the period of revolution for this orbit is twice the period for circular orbit *1*. Similarly, for orbit *3* the initial conditions ensure that the period is two times the period for orbit *2* and four times the period for circular orbit *1*. For orbit *4* the semimajor axis is four times the radius of the circular orbit. Consequently, Kepler's third law predicts that the period of revolution for this orbit equals eight periods for the circular orbit.

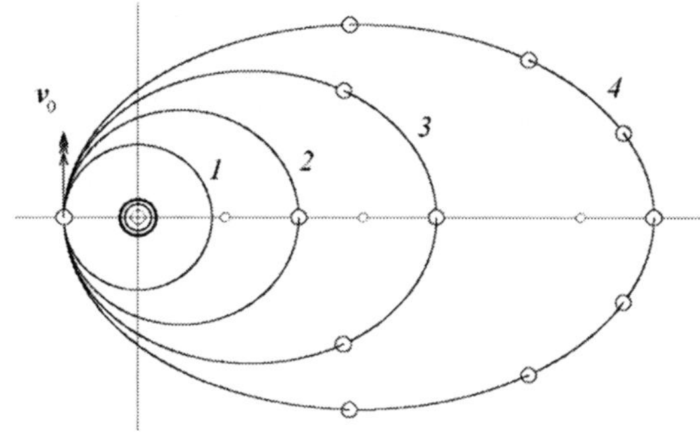

**Figure 2.4.** Elliptical orbits with common initial position and periods of revolution that relate as integer numbers 1 : 2 : 4 : 8.

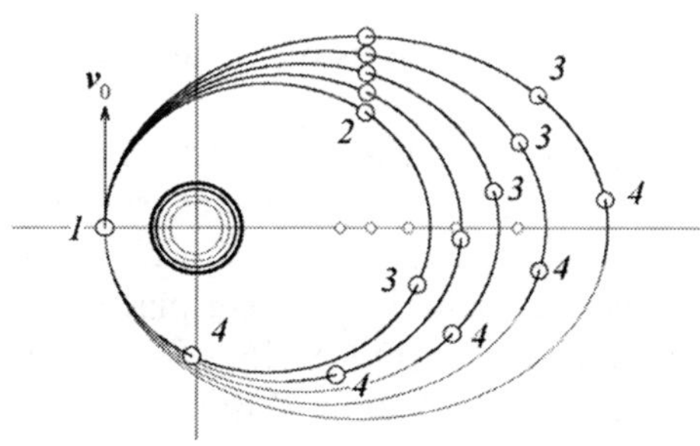

**Figure 2.5.** Evolution of the satellites' configuration after a simultaneous launch with slightly different initial velocities.

Observing the motion of the system after the simultaneous launch, we notice that the initial configuration for two satellites *1* and *2* is reproduced after every two revolutions of the satellite around the circular orbit. For three satellites *1*, *2* and *3*, the initial configuration is reproduced after four revolutions, and for the whole system after eight revolutions around the circular orbit. If the option 'Time marks' is chosen, spatial positions of the satellites are fixed on the screen at equal intervals of time after each revolution around the standard circular orbit (see the circles along the orbits in figure 2.4). The program can also plot the theoretical elliptical orbits if that option is chosen. These ellipses include the positions of the second foci, which are marked with small circles. To reproduce the simulation described above, as well as those described below, either the appropriate data can be entered or the corresponding example in the menu item 'Examples' can be chosen.

As we increase the value of the initial velocity, the elliptical orbit gradually elongates. When the initial velocity approaches the escape velocity, even a very small increment in the initial velocity dramatically increases the major axis and the period of revolution. We can observe how small variations in the initial velocity influence the orbit if we generate several long-period orbits that are produced by slightly different initial velocities and launch the satellites simultaneously. Figure 2.5 illustrates the evolution of the system. During the early stage of the motion, the positions of all the satellites are very close, but they become considerably more distant in the

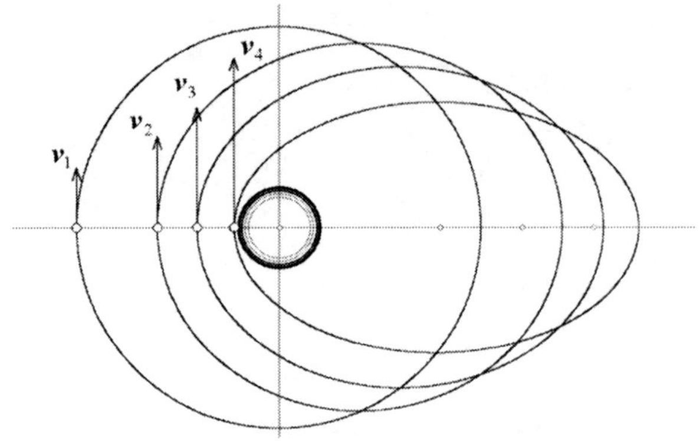


**Figure 2.6.** Orbits with equal major axes and equal periods of revolution.

course of time. The simultaneous positions of the different satellites in figure 2.5 are marked by the same numbers.

The program provides other ways to vary the conditions of the simulation. Choosing the option 'Arbitrary orbits' found in the menu item 'Orbits,' we can call up several orbits with arbitrarily chosen major axes and eccentricities. When we do so, the initial positions of the satellites may be found in different places for different orbits.

For example, it is possible to generate orbits with the same major axes but different eccentricities (figure 2.6). By Kepler's third law, the period of revolution depends only on the major axis of the elliptical orbit, regardless of its eccentricity. Therefore all such orbits in this example have equal periods of revolution. If all the satellites are launched simultaneously, they all return to their starting positions at the same moment, thus reproducing the initial configuration after every revolution.

The general idea of orbital motion in a planetary system can be illustrated if we generate several circular orbits with different radii. The larger the radius of the orbit, the greater the period of revolution and the smaller the orbital velocity (in sharp contrast with the case of a rotating solid, in which the velocities of particles increase in proportion to their distances from the center of rotation). If orbits with multiple periods are chosen that relate as small integers (say, 1 : 3 : 6 : 12), and if the initial positions of the planets are chosen to lie along the same radial line, we can observe the periodic reproduction of this unusual configuration of the system in which all the planets again occur simultaneously along the same radial line (a 'parade of planets').

## Questions and problems

1. **Third law for circular orbits.** The radii of two circular orbits relate as 4 : 1. Applying Newton's second law to circular motion in the central gravitational field, find the ratio of the periods of revolution for these orbits in terms of their radii. In other words, prove Kepler's third law for the special case of circular orbits. What is the ratio of the orbital velocities in this case?
2. (*) **Period of revolution and the initial velocity.** With what value of the transverse initial velocity (in units of the circular velocity for a given initial position) must a satellite be launched to get an elliptical orbit for which the period of revolution of the

satellite is twice the period for the circular orbit in which the satellite starts from the same position? What initial velocity provides the elliptical orbit shown in figure 2.3 (with a period of revolution four times the period for the circular orbit)? Verify your calculated values of the initial velocity in a simulation experiment.

3. **Linear dimensions of an orbit and the period of revolution.** By how much must the linear dimensions of an elliptical orbit be increased to obtain a period of revolution eight times greater than that for the original orbit? Does the answer depend on the eccentricity of the orbit? Verify your answer in the simulation experiment.
4. (*) **Mass and the period of revolution.** How does the period of revolution of a satellite depend on the mass of the planet? What is the ratio of the periods of two satellites that revolve about planets whose masses are in the ratio 4 : 1 and in orbits whose linear dimensions are the same?
5. (**) **Free-fall toward the Sun.** How long would the free-fall of a body toward the Sun be if the body starts to fall (with an initial velocity of zero) from the initial distance equal to the radius of the Earth's (nearly circular) orbit? During what part of this time would the body cover the first half of this distance?

## 2.4 The approximate nature of Kepler's laws

The orbital motion of any planet around the Sun is governed mainly by the force of gravitational attraction to the Sun. Actually, Kepler's laws are only an approximation to the planetary motions because they follow from Newton's laws only if the motion occurs under the single force that is directed toward a central fixed point and if this force decreases as the inverse square of the distance. In actuality, however, the Sun, which serves for a planet as the source of the major force, is not fixed but experiences small accelerations because of the other planets. Furthermore, the planets attract one another, so that the total force on a planet is not just that due to the Sun: other planets perturb the elliptical motion that would have occurred for a particular planet if that planet had been the only one orbiting an isolated Sun. Kepler's laws therefore are only approximate. Gravitational perturbations from other planets cause deviations from these simple laws. For example, the components of a double planet (like the Earth with the Moon) move along very complicated trajectories while orbiting the Sun.

That Kepler's laws are such good approximations to the actual planetary motions results from the fact that all the planetary masses are very small compared to that of the Sun. It is then convenient to treat the motion of a particular planet as slightly perturbed Keplerian motion and to determine the changes in the parameters of the ellipse that result from the small forces as time progresses. One of the famous achievements of celestial mechanics based on the perturbation technique was the discovery of Neptune from its perturbations of the motion of Uranus. Neptune was discovered in 1846 almost precisely where it had been predicted to be on the sky. The power of classical dynamics to explain and predict phenomena was triumphantly confirmed. Throughout history, the motion of the planets in the Solar System has served as a wonderful laboratory to confirm celestial mechanics in particular and classical mechanics in general.

Newton's conception of gravitation held firm until the beginning of the 20th century, when the notion of instantaneous action at a distance, implied in the law of universal gravitation, was recognized as generally inconsistent from the viewpoint of the special theory of relativity. In his general theory of relativity, Albert Einstein developed a wholly new, relativistic conception of gravitation, according to which the presence of matter, instead of creating a force field, causes a curvature of the four-dimensional space–time continuum, producing thus a universe in which free bodies travel through the curved space–time in geodesics (shortest four-dimensional paths). Einstein's approach to gravitation is based on the principle of equivalence according to which there is no way in principle to distinguish between a body in the reference frame that is subjected to a uniform acceleration in the absence of gravitation and the body in a reference frame that is stationary in a gravitational field.

The relativistic theory of gravitation yielded predictions of several phenomena that Newtonian theory cannot explain. These predictions include the bending of a ray of light passing near a very massive object such as the Sun, the red shift of light emitted near a very massive object and the slowing down of a clock in a strong gravitational field. These phenomena have been confirmed experimentally to the limits of observational accuracy. The new theory was also able to account for the additional precession of the orbit of Mercury about the Sun of 43 arc seconds per century, which remained unexplained in Newtonian analysis when all possible perturbations from other planets were taken into account. Even today, Newton's theory is of sufficient accuracy for almost all applications of celestial mechanics and astrodynamics. The major significance of Einstein's theory is its radical conceptual departure from classical theory and its implications for further growth in physical thought.

## Questions and problems

1. **Shapes of planetary orbits.** Which physical laws must be used for an exhaustive theoretical explanation of the actual (circular or elliptic) shapes of planetary orbits?
2. **Closed planetary orbits.** Which peculiarity of the law of universal gravitation is responsible for the closed shapes of orbits? Specify requirements (physical conditions) under which the orbit of a celestial body would be exactly elliptical.
3. **Properties of Keplerian motions.** The first Kepler's law claims that a body under the central inverse square gravitational force traces an elliptical orbit. What are the properties of these motions that are described by the second and third Kepler's laws?
4. (*) **Planetary orbits.** Which peculiarities in the construction of the solar system are responsible for the (approximately) elliptical shape of planetary orbits? Why are the real orbits only approximately elliptical?
5. (**) **Kepler's Laws for satellites.** The satellite that orbits the Earth is subjected to the gravitational forces exerted not only by the Earth, but also by the Sun. Explain why it is possible to apply Kepler's Laws to the motion of a satellite around the Earth as if this motion were governed solely by the gravitational force of the Earth. In your answer take into account that the gravitational pull of the Sun exerted on the satellite can be comparable with the pull of the Earth.

# Chapter 3

# Hodograph of the velocity vector for Keplerian motion

One of the most interesting aspects of Keplerian motion is concerned with the shape of its trajectory in velocity space. It occurs that if the body traces an elliptical orbit in space (a planet moving around the star, or a satellite of a planet), its trajectory in the velocity space is a circle. For an infinite parabolic or hyperbolic motion in a Newtonian central gravitational field, the hodograph of the velocity vector is also circular.

The vector of velocity of a moving body at any moment is directed *tangentially* to the spatial trajectory and so in curvilinear motion the direction of the velocity vector continuously changes. We obtain the trajectory of motion in *velocity space* as follows: for each point on the spatial trajectory, we draw the corresponding vector of velocity so that its tail lies at the origin of velocity space and its direction is parallel to the tangent to the spatial trajectory at the point in question. During the curvilinear non-uniform motion of the body, the direction and magnitude of this vector change. The end of this varying velocity vector generates a curve in velocity space. This curve is called the *hodograph* of the velocity vector.

## 3.1 Hodograph of the velocity for closed orbits

For a circular Keplerian orbit, the magnitude of the velocity is constant and so the variation of the velocity vector is reduced to a uniform rotation about the origin of velocity space. It is evident that the hodograph of the velocity vector for a circular Keplerian motion is itself circle whose center is located at the origin of velocity space. The radius of this circle equals the constant value of the circular velocity.

As a planet or satellite moves in an elliptical orbit, rotation of the velocity vector is non-uniform, and both the direction and magnitude of the vector change. However, these variations occur in such a way that the end of the velocity vector in this case also generates a circle in velocity space but the center which is not at the origin. In other words, the hodograph of the velocity vector for an arbitrary Keplerian

doi:10.1088/978-0-750-31100-7ch3  

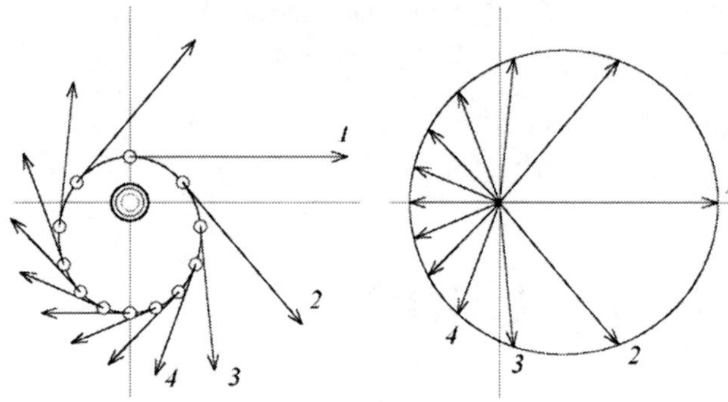


**Figure 3.1.** Keplerian orbit of a satellite and the velocity vectors in space (left) and hodograph of the velocity vector in velocity space (right). Coinciding numbers in the left and right sides refer to the same instants of time.

motion is a circle. An analytic proof of this property based on Newton's laws of motion can be found in section 11.4.

To illustrate this property of Keplerian motion, the simulation program numerically calculates the motion of a satellite (or a planet) in an elliptical orbit and simultaneously draws both the spatial trajectory in the left window on the screen and the hodograph of the velocity vector in the right window (figure 3.1). In equal time intervals, the vectors of velocity are fixed as tangents to the spatial trajectory shown in the left window and the same fixed vectors are shown in the velocity space. We can see clearly that during the motion of the satellite along the elliptical orbit, the end of the varying vector of velocity generates a circle in velocity space.

The lower semicircle of the hodograph (starting at point *1* and moving clockwise) corresponds to the first half of the elliptical orbit, during which the satellite moves clockwise from perigee toward the apogee and its speed decreases. The other (upper in figure 3.1) semicircle of the hodograph corresponds to the second part of the orbit, during which the satellite moves clockwise from apogee to perigee and its speed increases.

For an elliptical orbit, the diameter of this circular hodograph equals the sum of the magnitudes of the velocities at perigee and apogee. (We remember that at these points of an elliptical orbit the velocity vectors are oppositely directed.) The center of this circular hodograph corresponding to the elliptic motion is displaced from the origin of velocity space in the direction of the velocity vector at perigee through the distance equal to half the difference of the velocity magnitudes at perigee and apogee of the orbit (see figure 3.1).

The simulation experiment displays the orbital motion of a planet or a satellite and the corresponding behavior of the velocity vector in some time scale. Therefore, observing the simulation, we can see how the velocity of Keplerian motion changes with time. Positions of the satellite in the orbit and the corresponding vectors of the velocity are fixed in the simulation at equal time intervals. Consequently, even from the final static picture on the screen (see figure 3.1) we can judge (by comparing the angles between sequential positions of the velocity vector) variations in the angular velocity of rotation of the velocity vector during the Keplerian motion.

When starting the simulation experiment, we can vary the magnitude of the initial velocity. If we enter a value smaller than unity (that is, smaller than the velocity producing the circular orbit), the initial position will be the apogee of the elliptical trajectory. In this case the center of the circular hodograph will be displaced to the left from the origin of velocity space. The velocity vector at the beginning of the motion will be small in magnitude and will rotate slowly. As the satellite approaches the perigee of its orbit, the velocity vector rapidly increases, and its rotation becomes faster. Then as the velocity vector generates the upper semicircle of the hodograph, the variations occur in the reverse order.

## 3.2 Hodograph of the velocity for open orbits

The parabolic trajectory for Keplerian motion (the total energy for which is zero) can be considered the limiting case of motion along a strongly prolate elliptical orbit whose apogee approaches infinity. In this case, the velocity of the satellite at apogee approaches zero. The hodograph of the velocity vector for this limiting case of a parabolic motion is shown on the right side of figure 3.2, and the corresponding parabolic trajectory on the left. This hodograph is a closed circle whose diameter equals the velocity of the body at the vertex of the parabola (the point on the trajectory nearest to the center of force). The circular hodograph corresponding to the parabolic Keplerian motion passes through the origin of velocity space. The point of the hodograph which coincides with the origin corresponds to the infinitely remote point on the trajectory at which the velocity of the satellite is zero.

If a body, at some point of a Newtonian central gravitational field, gets an initial velocity in the transverse direction (perpendicular to the initial radius vector) and its magnitude equals the value of the escape velocity for this point, the vector of velocity in velocity space generates a semicircle (lying below the abscissa axis in the right side of figure 3.2), whose horizontal diameter is the vector of the initial velocity. Starting from this diameter, the velocity vector, with its end moving along the semicircle, eventually contracts into a point at the origin of velocity space as the body recedes to infinity. The semicircle is generated over an infinitely long time.

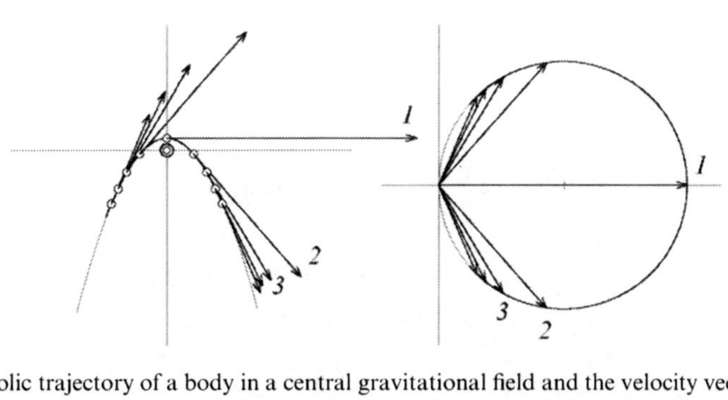


**Figure 3.2.** Parabolic trajectory of a body in a central gravitational field and the velocity vectors in space (left) and hodograph of the velocity vector in velocity space (right). Coinciding numbers in the left and right panels refer to the same instants of time.

To obtain the other half of the parabolic trajectory and the corresponding semicircle of the hodograph of the velocity vector (lying over the abscissa axis in figure 3.2), the program also simulates the motion of the body from infinity toward the vertex of the parabola. This simulation is achieved in the following way. When the body recedes far enough from the initial position (beyond the space region displayed on the screen), the program reverses the sign of one of its coordinates (namely, the sign of the coordinate measured from the initial position in the direction perpendicular to the initial radius vector) and simultaneously reverses the other component of the velocity. Hence, the further motion of the body occurs toward the initial position along the other half of the same parabola.

For an infinite hyperbolic motion in a Newtonian central gravitational field, the hodograph of the velocity vector is also circular (figure 3.3). In this case the radius of the velocity hodograph is smaller than the maximal velocity $v_P$ of the body at the point nearest to the center of force. The origin of velocity space is outside the circle.

In a hyperbolic motion, the body approaches the center of force from infinity, where its velocity is non-zero and is directed along one of the asymptotes of the hyperbola. In the hodograph, the velocity at infinity is tangent to the circle. In the course of motion the velocity vector generates a part of the circle away from the origin: starting from the point of tangency, the magnitude of the vector gradually increases and reaches its maximal value when the vector extends from the origin of velocity space to the farthest point of the circle. This vector is the velocity of the body when it is at the vertex of the hyperbolic trajectory (the point nearest to the force center).

The velocity vector then gradually shortens, its end moving further along the circle. Eventually the end of the velocity vector reaches the other point of tangency where the velocity assumes its initial magnitude and the body is infinitely remote. This second tangent to the hodograph is directed along the other asymptote of the hyperbolic trajectory, along which the body recedes.

In the simulation, the motion begins from orbital position nearest to the center of force, where the initial velocity of the body is transverse. (When entering data, we can vary the magnitude of the initial velocity through a limited range.) The body

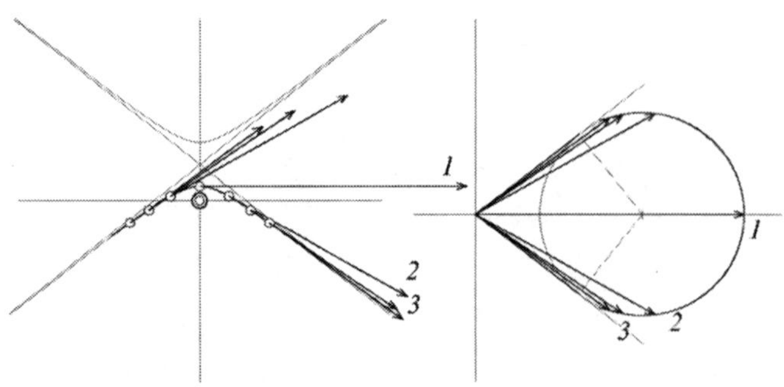


**Figure 3.3.** Hyperbolic trajectory of a body in a central gravitational field and the velocity vectors in space (left) and hodograph of the velocity vector in velocity space (right). Coinciding numbers in the left and right panels refer to the same instants of time.

moves from the initial position to infinity along a half of the hyperbola. In velocity space, the vector of velocity generates the lower part of the circular hodograph. In figure 3.3 the end of the velocity vector moves along the circular path from point *1* to point *2*, then to point *3* and further on, approaching the final point (point of tangency) asymptotically during an indefinitely long time.

To show the other part of the hodograph, the program also simulates the motion from infinity toward the initial position along the left part of the hyperbolic trajectory. This is done in the same way as for the parabolic motion described above.

## Questions and problems

1. (*) **Radius of the hodograph of the velocity vector.** For any Keplerian motion, the trajectory in velocity space, that is, the hodograph of the velocity vector, is a circle (or a part of a circle for an infinite hyperbolic motion). Imagine a satellite orbiting the Earth in an elliptical orbit. Let the velocity of the satellite at the perigee be 1.25 times the circular velocity for this point. Calculate the radius of the velocity hodograph for the motion in this elliptical orbit. Express this radius in units of the circular velocity for the perigee of the elliptical orbit. Calculate also the coordinates of the center of this circular hodograph in velocity space (position of the center relative to the origin of velocity space).
2. (*) **Velocity at the end of the minor axis.** Taking into account the circular shape of the velocity hodograph, express the satellite's velocity at the end of the minor axis in terms of the velocities $v_p$ and $v_a$ at perigee and apogee of its elliptical orbit.
3. (*) **Velocity hodograph for a hyperbolic trajectory.** Consider a celestial body (say, comet) that approaches the Sun from infinity, passes through the perihelion of its hyperbolic trajectory with the velocity equal to 1.6 times the circular velocity for this point, and then recedes to infinity along the other asymptote of the hyperbola. Calculate the radius of the velocity hodograph for this hyperbolic motion (in units of the circular velocity for the perihelion). Find also the position (coordinates) of the center for this circular hodograph in velocity space. During the motion, the end of the velocity vector only traces part of the circle. Calculate the central angle that subtends this arc of the circular hodograph.
4. (**) **The circular form of the velocity hodograph.** Try to prove analytically (on the basis of Newton's laws of motion) that for any motion in the Newtonian central gravitational field the trajectory in velocity space is a circle (or an arc of a circle).

# Chapter 4

## Orbits of satellites and trajectories of missiles

In the preceding programs, devoted to the simulation of motion of a body in a Newtonian inverse-square central gravitational field, it is possible to choose the height of the initial position and the magnitude of the initial velocity of the satellite arbitrarily. However, it is not possible to change the direction of the initial velocity. The initial velocity is always directed horizontally (perpendicularly to the local vertical line, i.e. transverse to the radius vector). For such conditions of the launch, the major axis of the orbit is oriented along the vertical line passing through the center of the Earth and the initial position, and this initial position is either the perigee or the apogee of the orbit depending on the magnitude of the initial velocity.

The program 'Missiles and Satellites' allows us to also vary the direction of the initial velocity. The program simulates the motion of an artificial satellite orbiting the Earth (or of a ballistic missile, if the trajectory intersects the Earth's surface) during the passive (freely falling) part of its trajectory. We can vary the conditions for launching the satellite or missile by choosing different values of the initial position and the initial velocity for the passive motion of the projectile in the Earth's gravitational field. It is also possible to set the initial conditions for several missiles or satellites and observe their motion either simultaneously or sequentially.

We emphasize that in the program we specify the initial position and the initial velocity for the *passive* orbital motion of a ballistic projectile or a satellite. During the lift-off, a missile or a spacecraft gains speed because of the thrust created by its jet engines. During this so-called *active stage* of the flight, we are dealing with the jet propulsion of a rocket projectile rather than with the free-fall of an artificial celestial body. The initial position which we indicate in the program corresponds to the end point of the active stage. At this point the jet engine of the last stage of the rocket terminates its operation. The initial velocity for the orbital or ballistic flight is determined by the energy delivered to the projectile during the active stage of its flight.

The further passive motion of the projectile occurs only under the force of Earth's gravitation and probably under the force of air resistance. To simulate the motion of

doi:10.1088/978-0-750-31100-7ch4

a projectile subjected to atmospheric resistance, we choose the corresponding option in the menu and enter additional parameters that characterize the atmosphere of the planet and the cross-sectional area and mass of the spacecraft or the missile. However, it is reasonable to begin with an investigation of motion in cases for which air resistance is totally absent or insignificant.

## 4.1 Families of Keplerian orbits

### 4.1.1 Orbits with various directions of the initial velocities

How does the orbit of a satellite change if we vary the direction of the initial velocity? Figure 4.1 shows several orbits of satellites launched from the same initial position $S$ with initial velocities that are equal in magnitude but different in direction. At the initial position $S$, the satellites have the same value of potential energy (if we assume for simplicity that they have equal masses) and the same value of kinetic energy, since their speeds are equal. Consequently these orbits correspond to the same value of total energy and so have equal major axes and equal periods of revolution. An analytic proof of these statements, based on the laws of motion, can be found in section 11.3. With this computer simulation, we can verify these properties experimentally.

In figure 4.1, vector $\mathbf{v}_0$ is directed horizontally and, since its magnitude is greater than the circular velocity $v_c$, the initial position $S$ is the perigee of the corresponding orbit, as we discussed earlier. The eccentricity $e_0$ of this orbit can be calculated using the following simple formula:

$$e_0 = \frac{v_0^2}{v_c^2} - 1. \tag{4.1}$$

A proof of this expression is given in section 10.5, where the geometric properties of Keplerian orbits are discussed in greater detail.

For any other direction of the initial velocity, the elliptical orbit is characterized by a greater eccentricity $e_1$ than the preceding orbit with the horizontal initial velocity ($e_1 > e_0$). If the initial velocity $\mathbf{v}_1$ makes an angle $\theta$ with the local vertical, the

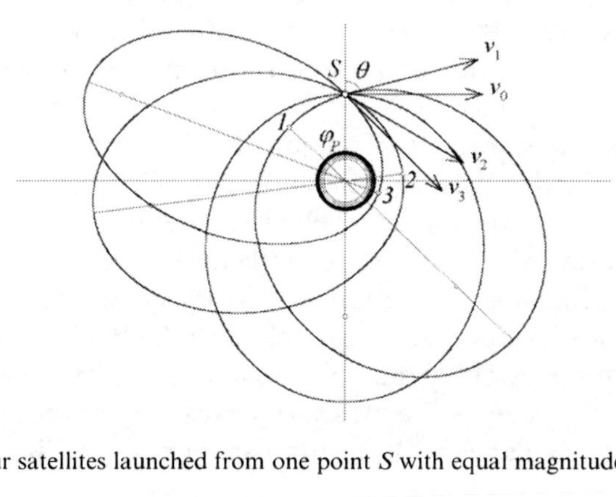


**Figure 4.1.** Orbits of four satellites launched from one point $S$ with equal magnitudes, but different directions of the initial velocities.

eccentricity $e_1$ of the orbit can be expressed in terms of the eccentricity $e_0$ of the orbit which corresponds to the horizontal direction of the same in magnitude initial velocity:

$$e_1^2 = e_0^2 + \left(1 - e_0^2\right)\cos^2\theta. \qquad (4.2)$$

If we change only the direction of the initial velocity (preserving its magnitude), the length of the major axis of the resulting elliptical orbit remains the same. Since for a horizontal initial velocity that is greater than the circular velocity ($v_0 > v_c$) the perigee of the orbit occurs at the initial position, for any other direction of the initial velocity the perigee is lower than the initial position. Indeed, for a given major axis, the greater the eccentricity the lower the perigee.

If the initial velocity is directed slightly upward, like vector $\mathbf{v}_1$ in figure 4.1 (that is, if the velocity forms an acute angle $\theta$ with the local vertical line), the perigee $P$ of the new orbit (point *1* in figure 4.1) is displaced from the initial position through an angle $\varphi_P$ in the direction opposite to the initial velocity. If we change the direction of the initial velocity downward through some angle from the horizontal direction, like vectors $\mathbf{v}_2$ and $\mathbf{v}_3$ in figure 4.1, the perigee of the orbit is displaced forward in the direction of the initial velocity (points *2* and *3*, respectively).

For the orbit that corresponds to an angle $\theta$ between the initial velocity and the vertical line, the angular position of the perigee (angle $\varphi_P$ between the radii of the initial position $S$ and the perigee $P$) can be calculated by the formula

$$\tan\varphi_P = -\frac{\sin\theta\cos\theta}{(v_c/v_0)^2 - \sin^2\theta}, \qquad (4.3)$$

where $v_c$ is the value of the circular velocity for the initial position.

After launching a body into orbit, we obtain an Earth satellite only if the distance of the perigee is greater than the Earth's radius. Otherwise the missile travels along the ellipse until it hits the ground at the point where the ellipse intersects the surface of the Earth. In such cases we call the body a ballistic missile rather than a satellite.

The latter situation is characteristic of all launches for which the passive orbital motion starts almost from the Earth's surface. In other words, it is impossible to launch a satellite directly from the ground (even in the absence of air resistance), since the orbit of such a projectile inevitably crosses the Earth's surface. Examples of two ballistic trajectories are shown in figure 4.2. The projectiles are launched from one point $S$ of the Earth's surface with different velocities $\mathbf{v}_1$ and $\mathbf{v}_2$, both being greater in magnitude than the circular velocity. The greater the initial velocity, the higher the apogee of the orbit, but the projectile nevertheless hits the ground, a result that is independent of how great the initial velocity is (provided, of course, that the initial velocity does not exceed the escape velocity). Only if the velocity at the initial position is directed horizontally and is greater in magnitude than the circular velocity, does the orbit graze the Earth's surface (exactly at the starting point) rather than intersect it. For a planet with an atmosphere, a satellite with such an orbit is impossible.

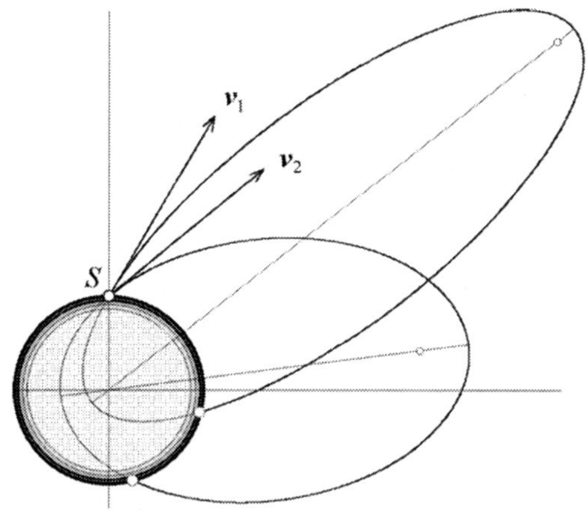


**Figure 4.2.** Two ballistic trajectories of missiles launched from one point of the Earth's surface with different initial velocities.

For the initial velocity directed exactly downward, the trajectory is clearly the segment of the vertical line starting from the initial position and ending at the ground. However, we can consider the body before hitting the ground to be moving along a trajectory that is the limiting case of more and more flattened ellipses with one of the foci at the center of the Earth. Eventually, such a flattened ellipse with its eccentricity $e \to 1$ degenerates into a rectilinear segment. The foci of this degenerate ellipse coincide with the ends of the segment. One of its ends is located at the Earth's center.

If the initial velocity of the missile is directed vertically upward, the trajectory is a part of the same degenerate ellipse. The second focus of the degenerate ellipse coincides with the highest point of this rectilinear trajectory.

Several ballistic trajectories of missiles are shown in figure 4.3. Each of these orbits starts from the same point on the Earth's surface, but the initial velocities point in different directions. Their magnitudes are equal to that for a circular orbit whose radius is the Earth's radius (that is, for a hypothetical extremely low orbit around the Earth). We note the following interesting properties of this family of trajectories.

1. The major axes of the ellipses are equal since the orbits correspond to satellites with equal total energy per unit mass.
2. The major axis of each ellipse is oriented parallel to the corresponding vector of the initial velocity.
3. One of the foci of each ellipse is located at the center of the Earth, while the other focus lies on the circle whose center is at the initial position and whose radius equals the Earth's radius.

If the corresponding option is chosen, the theoretically predicted elliptical trajectories (the thin lines in figure 4.3) are displayed by the program before the simulation. The actual trajectory, calculated by a numerical integration of the

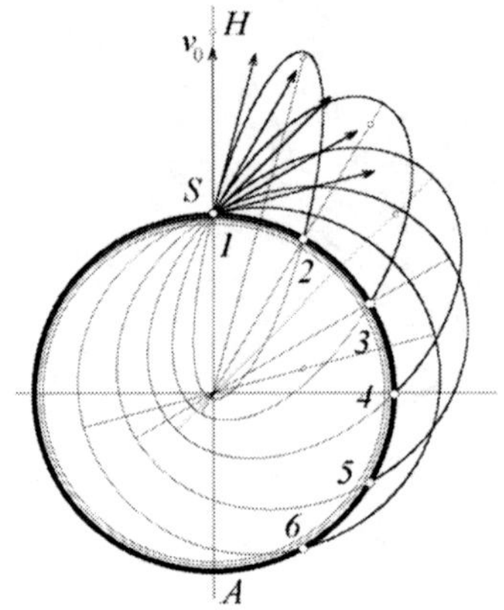


**Figure 4.3.** Ballistic trajectories of missiles launched in different directions from the same point *S* on the Earth's surface, with initial velocities whose magnitudes are equal to the circular velocity.

equations of motion, is traced on the screen in a different color (the thick curves in figure 4.3). In the absence of an atmosphere, the trajectories of the actual motion coincide exactly with the predicted elliptical curves until they cross the Earth's surface. We note how the point at which a projectile hits the ground recedes from the starting point *1* as the angle between the vertical line and the initial velocity is increased from zero to $\pi/2$. For the idealized airless case, this point of incidence tends to the antipodal point (the opposite end of the diameter passing through the initial position) as the direction of the initial velocity approaches the horizontal. In space, trajectories of this set occupy a particular region over the Earth's surface. This axially symmetric region is bounded by a surface of rotation which is an ellipsoid whose foci are located at the Earth's center and the initial position *S*. The major axis of the ellipsoid joins the antipodal point *A* and the highest point *H* (see figure 4.3). This surface is discussed in greater detail in the following section 4.1.2.

### 4.1.2 Satellites with equal magnitudes of the initial velocities

Imagine a rocket, launched from the Earth, rising vertically, and at the highest point of its flight exploding into many fragments that fly off in all directions with equal speeds. The further motion of the fragments occurs under the action of the central force of the Earth's gravity. Thus, they become the Earth's satellites, orbiting along various elliptical Keplerian orbits.

The motion of such fragments can be simulated with the program `MISSLES AND SATELLITES`. We enter the height of the initial position and the common magnitude of the initial velocity that each fragment receives in the explosion. (The initial height and the initial speed can be expressed either in kilometers and kilometers per second, respectively, or in the natural units of the Earth's radius and the circular velocity for the height of the initial position.) The magnitude of the initial velocity should not exceed that of the escape velocity. Then several orbits with different values for the direction of the initial velocities can be prescribed. (If desired, this prescription can be chosen to be the corresponding example from the item

'Examples' in the menu.) Figures 4.4 and 4.5 are illustrations of the family of elliptical trajectories of bodies starting from a common point $S$ with equal speeds and different directions. For the family of orbits in figure 4.4 the initial velocity of the satellites is greater in magnitude than the circular velocity for the initial position $S$ and in figure 4.5 it is smaller.

One of the foci is common to each of the elliptical orbits. It is located at the center of the Earth. The second focus of each orbit lies on a circle whose center is located at the common initial position $S$. This circle is shown by a dashed line in figures 4.4 and 4.5. Its radius equals the distance between the initial position $S$ and the highest point $N$ reached by the fragment that moves vertically upward from the initial position $S$. Since the magnitudes of the initial velocities of each of the fragments are equal, and since the motion of each fragment begins at the same point, the total energies of each are equal. Therefore, the major axes of the orbits are the same and, by Kepler's third law, so also are the periods of revolution. That is, all the fragments

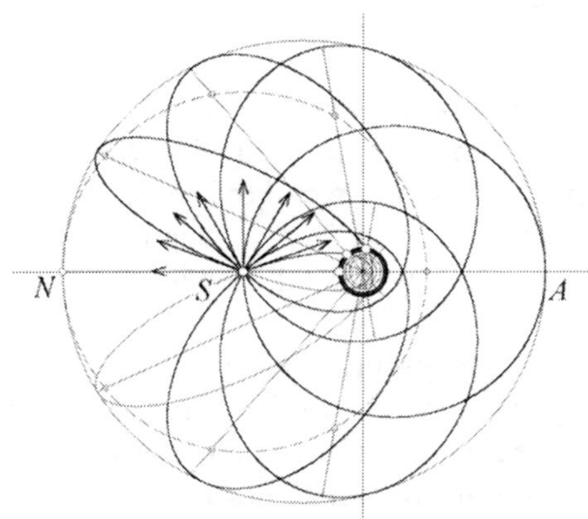


**Figure 4.4.** A set of elliptical orbits of fragments scattered from the same initial position $S$ in different directions with equal speeds. This speed exceeds that which generates a circular orbit ($v_0 > v_c$).

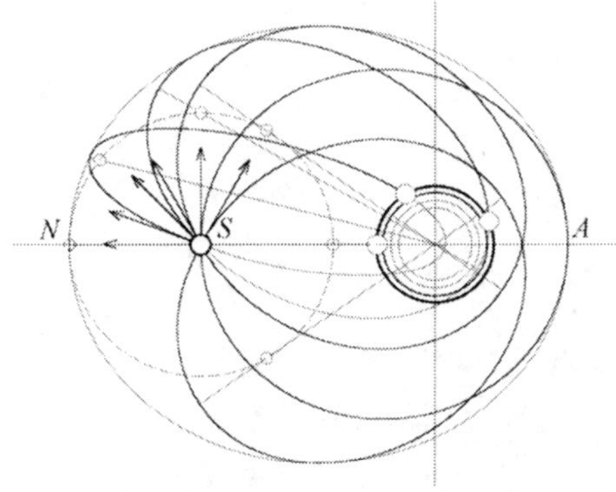


**Figure 4.5.** Elliptical orbits and ballistic trajectories of fragments scattered from one initial position $S$ for the case in which $v_0 < v_c$.

whose elliptical orbits do not intersect the Earth's surface simultaneously return to the initial position after a revolution.

These orbits are confined to a particular spatial region. Its boundary—the envelope for the family of orbits—is an axially symmetric surface whose axis of symmetry passes through the center of the Earth and the point at which the explosion occurs. A detailed analysis shows that it is a surface of revolution of an ellipse whose axis of revolution is the major axis and whose foci are at the center of the Earth and the initial position $S$. Diagrams of this ellipsoid are shown in figures 4.4 and 4.5 in which the enveloping surface (more precisely, its section by a plane) is depicted by a thin solid line. The dimensions and eccentricity of the ellipsoid are determined by the position of the initial point and by the magnitude of the initial velocities of the fragments. A rigorous derivation of the parameters of this bounding surface (based on the geometric properties of Keplerian orbits) is given in section 11.6.

### 4.1.3 Orbits of satellites launched in one direction with different magnitudes of the initial velocities

Next we consider the elliptical orbits traced by satellites that are launched from the same point in a Newtonian gravitational field and in the same direction, but at different speeds. Orbits like these were studied earlier for the case in which the velocity was directed horizontally. We recall that here the major axes of all the orbits are oriented along the vertical line passing through the initial position. One of the orbits is circular. It is generated by a satellite whose initial velocity equals the circular velocity.

For any other direction of the initial velocity, a circular orbit cannot be generated no matter what the magnitude of the initial velocity might be. Several orbits for which the initial velocities make the same acute angle with the upward vertical are shown in figure 4.6. The initial position $S$ is the only point common to the orbits. They are tangent to one another at this point, because here their velocity vectors lie in the same direction.

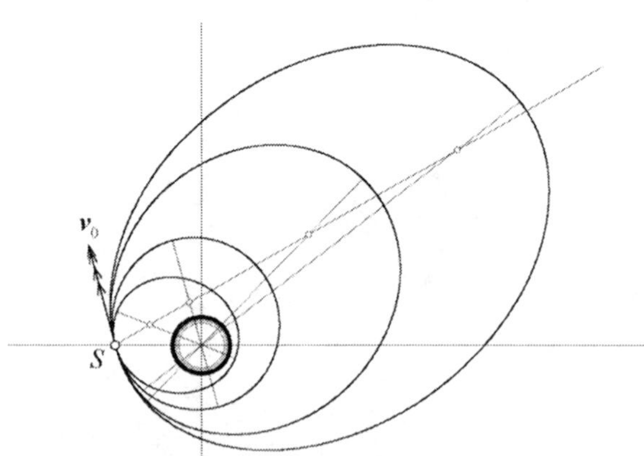


**Figure 4.6.** Elliptical trajectories of satellites launched in one direction with different magnitudes of the initial velocities.

An interesting property of this collection of elliptical orbits is related to the position of their foci. One focus is common to all the orbits. It is located at the center of the Earth and so of course this set of foci is represented by a single point. The foci of the other set are located on the one and the same straight line that passes through the initial position $S$ (see figure 4.6). This line forms an angle with the upward vertical at the initial position that is twice the angle formed by the initial velocity with the upward vertical. (For horizontal initial velocities, this property means that the foci of this second set lie on a downward vertical through the initial position.) This property of the set of orbits under consideration can be easily proved geometrically if we remember the optical property of the ellipse (see section 10.4), according to which all light rays emanating from one focus are reflected by the elliptical mirror toward the other focus.

Figure 4.7 shows a set of ballistic trajectories of missiles launched at different speeds from a point on the Earth's surface. The velocity of each is directed at an angle of 45° with the upward vertical. In the absence of air resistance, these trajectories are portions of ellipses with a common focus at the center of the Earth. If the initial velocity is small compared to the circular velocity, the portion of the ellipse above ground is approximately a parabola. This is the parabolic trajectory that we usually assign to a projectile in the approximation of a 'flat Earth' and in the absence of the air resistance.

For small initial velocities and consequently short ranges (compared to the Earth's radius), the gravitational field of the Earth can be considered to be uniform, i.e. constant in magnitude and direction (homogeneous) along the whole trajectory, and the above approximation is clearly applicable. However, we should keep in mind that actually the trajectory is a portion of an ellipse with one of the foci at the center of the Earth.

All the trajectories of the set are tangent at the initial position. Orientations of their major axes depend on the magnitude of the initial velocity. We note that the other foci of all the ellipses are located on the straight line that passes horizontally

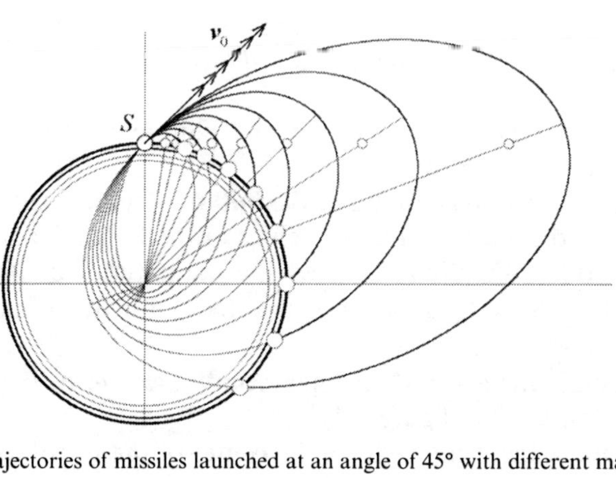


**Figure 4.7.** Ballistic trajectories of missiles launched at an angle of 45° with different magnitudes of the initial velocities.

through the initial position. As we have shown above, such an alignment of the second foci is explained by the optical property of the ellipse. The projectile whose initial velocity equals the circular velocity (also called the first cosmic velocity), hits the ground at a point whose angular distance from the initial position is exactly 90°.

## Questions and problems

1. **Launching a satellite with the help of a cannon.** Is it possible to launch an Earth satellite by shooting a ball from a cannon which remains on the ground? The cannon is powerful enough to provide the ball with a velocity exceeding the circular velocity.
2. **Orbits with the initial velocities equal to the circular velocity.** Consider several orbits of missiles starting from one and the same initial position on the surface of the Earth with various directions of the initial velocity. Let the magnitude of the initial velocity be equal to the circular velocity for the chosen initial position $S$ on the surface (that is, to the first cosmic velocity, figure 4.3). Assume that there is no atmosphere.

   (a) Prove that the major axis of such an orbit is parallel to the initial velocity vector. Prove also that the length of the semimajor axis equals the distance from the center of the Earth to the initial position $S$. (This distance is the radius $R$ of the Earth if $S$ is on the Earth's surface.)

   (b) Calculate the maximal height over the Earth's surface reached by one of the missiles (express your answer in units of the Earth radius). What initial angle provides this maximal height?

   (c) (*) Show that the second focus of these orbits is located on a circle whose center is at the initial position, and whose radius equals the Earth's radius. (The diameter of the circle is the line between the Earth's center and the point of maximal height).

   (d) (*) Which of the missiles is the last to hit the ground? How much time elapses between launching and striking the ground?

   (e) (**) Find the angular position of the point, measured from the initial position, at which the missile hits the ground if its initial velocity makes an angle $\theta$ with the vertical line at the initial position. How long does the flight of the missile last?

   (f) (*) What is the eccentricity of the elliptical bounding surface which envelopes all the trajectories?
3. **Set of orbits with equal magnitudes of the initial velocities.** Let the explosion of a rocket occur at a distance $r_0$ from the center of the Earth, let all the fragments be scattered in various directions, and let the magnitudes of the initial velocities be equal to $1.1 v_c$, where $v_c$ is the circular velocity for the initial position (for the distance $r_0$).

   (a) (*) Calculate the greatest distance from the center of the Earth reached by the fragments (point $N$ in figure 4.4).

   (b) (*) Calculate the distance from the center of the Earth for the most remote point reached by the fragments on the vertical line antipodal to the initial position of the explosion (point $A$ in figure 4.4). Which of the fragments reaches this point?

Adding $r_0$ to the calculated distance, we get the distance of this most remote point $A$ from the initial position (from the point of the explosion). Compare this distance with the distance of point $N$ from the center of the Earth calculated in the preceding item (a). Can you explain without calculations why these distances are equal?

(c) (*) Prove that the second foci of all the elliptical orbits of the set lies on a circle whose center is located at the common initial position $S$, and whose radius equals the distance between the initial position $S$ and the highest point $N$ reached by the fragment that moves along the vertical line passing through the initial position.

(d) (*) Calculate the eccentricity of the elliptical boundary that envelopes all the trajectories.

(e) (**) Prove that the boundary of the region occupied by the trajectories of the fragments is a dilated ellipsoid of rotation whose major axis is the segment joining the above mentioned points $N$ and $A$ (figure 4.4), and whose foci are located at the initial position $S$ of all the orbits and at the center of the Earth.

4. (*) **Orbits of satellites launched from one point in one direction with different magnitudes of the initial velocities.** Prove that the second focus of each orbit of the set is located on one straight line passing through the initial position (see figure 4.6). Prove that this line forms an angle with the upward vertical that equals twice the angle formed by the initial velocity with this vertical.
5. (*) **The ultimate height of rise and the flight time.** A ballistic missile is launched from the surface of the Earth with initial velocity directed vertically upward (away from the center of the planet). The missile rises rectilinearly with a deceleration caused by gravity, reaches some maximal altitude, and then starts to fall down along the same vertical line. Find the maximal distance of the missile from the center of the planet, if the initial velocity of the missile is equal to the circular velocity for the very low orbit (to first cosmic velocity). (***) Calculate the flight time, that is, the lapse of time from the start of the missile to the moment at which it hits the ground. Express your answer in units of the period of revolution along the extremely low circular orbit.
6. (***) **From the Pole to an equatorial target.** Calculate the minimal initial velocity which must be imparted to the ballistic missile on the North Pole of the Earth in order to hit the target located on the equator. Express the magnitude of required velocity in units of the circular velocity for an extremely low (hypothetical) orbit (first cosmic velocity). Find also the required direction of this velocity (the angle which this velocity should make with the horizon).

## 4.2 Evolution of an orbit in the atmosphere

Air resistance is added to the gravitational force whenever a satellite is launched into an orbit which passes through the atmosphere. The program `MISSILES AND SATELLITES` is able to display the evolution of the orbit under the influence of the atmosphere.

If the satellite's orbit passes over the Earth at large altitudes where the air is extremely rarefied, the force of air resistance experienced by the satellite is very small. However, the influence of the air resistance on the orbit is of a secular, accumulating (and increasing) character. During large time intervals, a very small atmospheric drag can produce significant variations in the satellite's orbit.

### 4.2.1 Evolution of an elongated elliptical orbit

In particular, if the orbit is initially an elongated ellipse whose perigee lies within the upper strata of the atmosphere, the apogee of the orbit becomes lower after each revolution, and the orbit gradually approaches a circle. To explain the evolution of the orbit, we must take into account that for a short time the satellite travels through the atmosphere each time it passes through the perigee. Because of air drag, the satellite emerges from the atmosphere with a speed slightly smaller than it had before entering.

As an approximation, we can consider that the satellite moves along a Keplerian ellipse under only the force of Earth's gravity and that it experiences air resistance almost instantaneously only at one point of its orbit, as if at the perigee it pierced an invisible thin wall. Consequently, each time the satellite passes through perigee, its speed is slightly reduced. This almost momentary reduction in speed at perigee influences first of all the distance of the apogee from the center of the Earth while leaving the distance of perigee very nearly the same. Thus, the orientation of the major axis of the orbit is almost unchanged, but the major axis itself diminishes with each revolution. The lower the perigee, the greater the effect.

Figure 4.8 shows the trajectories of two satellites launched into almost identical orbits. (The major axes of the orbits are oriented differently to make observation of the motion more convenient.) The only essential difference between the orbits is that initially the perigee of satellite *1* is slightly lower than that of *2*. The dashed circle around the planet indicates the (conventional) upper boundary of the atmosphere. Passing through the perigee, satellite *1* experiences a greater air resistance than does satellite *2* and the apogee of its orbit is lowered more rapidly. When satellite *1* hits

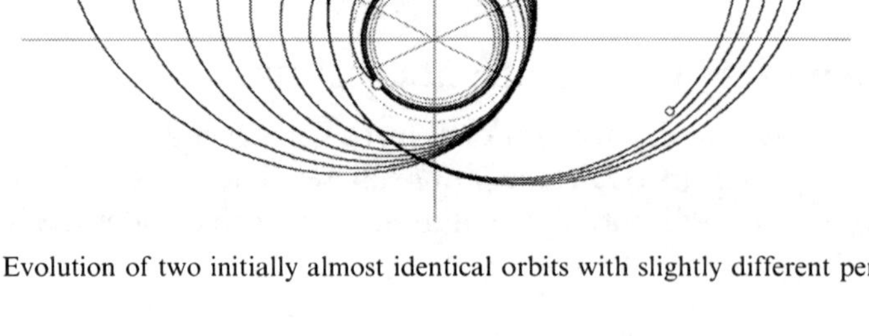


**Figure 4.8.** Evolution of two initially almost identical orbits with slightly different perigee height.

the ground, the other is still orbiting the Earth, though the apogee of its orbit is also gradually becoming lower.

In fact, air resistance acts along an extended part of the orbit in the vicinity of the perigee rather than at this single point. The result is that the height of the perigee is also decreased with each revolution, although the apogee height is decreased by a much greater amount because of air resistance experienced by the satellite near the perigee of its orbit. As the major axis of the orbit becomes smaller, the period of one revolution also diminishes. In addition the orbit gradually approaches a circle.

### 4.2.2 Late stage of the evolution and the aerodynamical paradox

When the orbit of the satellite becomes nearly circular, the satellite is subjected to air drag all along the orbit. Dissipation of the energy caused by air resistance causes the height of the satellite to diminish with each revolution. Its actual trajectory is a gradually twisting spiral. When the trajectory reaches the low strata of high atmospheric density, air resistance increases sharply, and the satellite cannot complete the next loop. Figure 4.9 depicts this final stage of evolution of the orbit. To make the changes in the trajectory easily observable, the effects of air resistance are exaggerated. Actually, for a planet with an Earth-like atmosphere, this spiraling trajectory descends at first rather slowly in almost circular loops.

Considering consecutive loops of the trajectory as circles with decreasing radii, we conclude that the orbital velocity of the satellite increases with each revolution. Strange as it may seem, it occurs that because of air resistance, the satellite is accelerated in the direction of its motion, as if the retarding force of the air resistance were pushing the satellite forward—a most unexpected result. This surprising consequence of the laws of dynamics is called the *aerodynamical paradox* of the satellite (or *satellite paradox*). Indeed, it looks paradoxical that the presence of air-drag force accelerates the satellite rather than decelerates it.

However, there is no real mystery in this phenomenon. Actually, of course, it is the force of gravity rather than the air drag that accelerates the satellite. The total acceleration is the vector sum of the gravitational acceleration, directed toward the

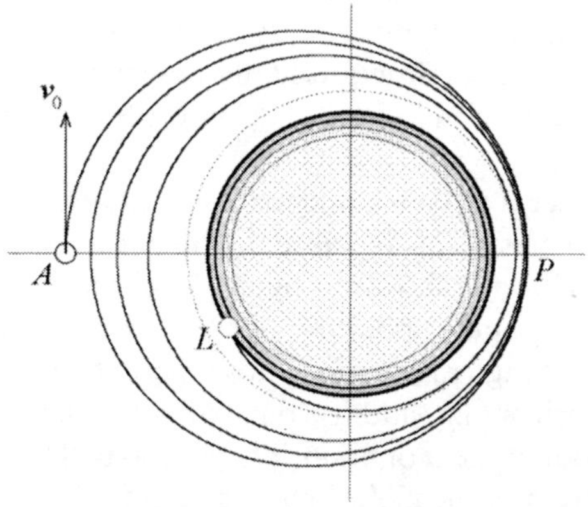


**Figure 4.9.** The late stage of evolution of an orbit in the atmosphere.

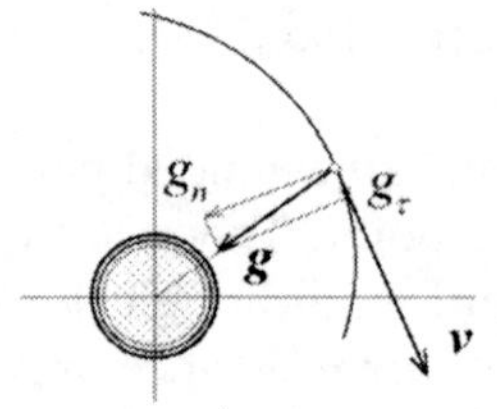


**Figure 4.10.** Components of the acceleration created by the gravitational force.

Earth's center, and the acceleration created by air resistance, directed opposite to the velocity. The actual motion of the satellite occurs along a twisting spiral rather than a circle. Consequently, the vector of the normal to the trajectory is directed not toward the center of the Earth but rather slightly behind the center. Hence, the gravitational force has a forward component along the trajectory.

Figure 4.10 shows the tangential component $g_\tau$ of the acceleration **g** created by the gravitational force. This tangential component points forward along the velocity vector **v**. It can be shown that the magnitude of this component is twice as great as the magnitude of the deceleration due to the air drag. Hence the total tangential acceleration is equal and opposite to that due to the air drag, and so the speed of the satellite increases.

The tangential acceleration occurs just as if the direction of the resistance of the air were reversed so that the air pushes the satellite forward! This unexpected effect is produced by the combined actions of the tangential air resistance and the radial gravitational force.

Thus, atmospheric resistance causes a gradual increase in the speed of the satellite instead of a gradual decrease. In the case of an elliptical orbit, this conclusion is true for the speed averaged over a complete revolution.

The increase of the speed does not contradict the law of the conservation of energy. Clearly, because of the loss of energy through friction with the air, the total mechanical energy diminishes in spite of the increasing velocity of the satellite. The resulting increment in the kinetic energy is more than canceled out by a greater decrease in the potential energy. Indeed, in the inverse-square gravitational field the mean value of the (negative) potential energy in magnitude is twice the mean value of the kinetic energy. As the satellite gradually descends, half of its loss in potential energy appears as an increase in kinetic energy, while the other half is dissipated as heat by the air-drag force.

Because of air resistance a missile launched horizontally, say, to the east, can hit the ground in the opposite hemisphere. We note that this result is not possible if there is no atmosphere. Such a situation is shown in figure 4.11, which resembles a famous, prophetic illustration from a popular book by Newton. Imagine that we position a large cannon at the top of a very high mountain, whose summit emerges over the atmosphere. We point the cannon horizontally and fire one ball after another. The first cannon ball falls to Earth near the foot of the mountain. The second ball has a greater initial velocity and bends round some part of the globe before falling to the ground. Finally, the ball is fired with a velocity that is sufficient to make the ball orbit around the Earth.

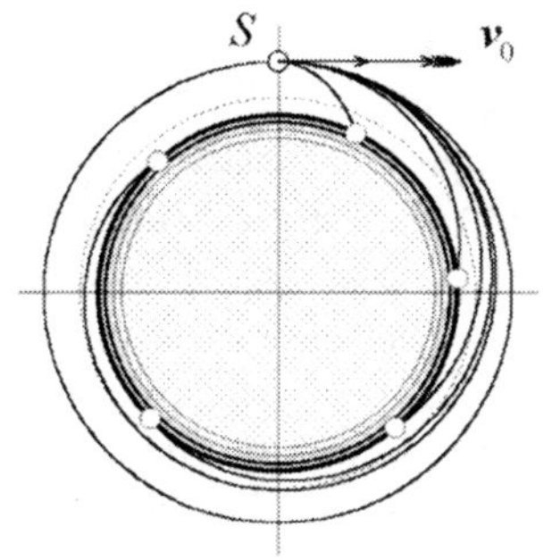


**Figure 4.11.** Trajectories of a ball fired horizontally by a cannon from a high mountain.

From this picture we can conclude that Newton was the first to build a bridge between familiar everyday phenomena on the Earth and the mysterious motions of the celestial bodies. Looking forward much further than his famous predecessors, Galileo and Kepler, Newton asserted that the motions of a cannon ball and of the heavenly bodies are governed by the same universal laws of physics. Moreover, we may guess from this picture that Newton may have foreseen that sooner or later man would launch an artificial satellite of the Earth!

### 4.2.3 Air density over the Earth

In the simulation program it is assumed that the resistance of the air is proportional to the local density $\rho$ of the atmosphere and to the square of the satellite's velocity, and is directed opposite to the velocity. These assumptions are certainly valid for objects like satellites in the atmosphere that move in a gas medium with velocities greater than the characteristic speed of thermal motion of the gas molecules.

The density $\rho$ of air is assumed in the program to decrease exponentially with the altitude $h$ above the surface of the planet:

$$\rho(h) = \rho_0 e^{-h/H}. \tag{4.4}$$

Here $\rho_0$ is the density at the surface (at $h = 0$) and $H$ is the characteristic height at which the density is $e \approx 2.72$ times smaller than at the surface. This expression for the dependence of the density on height $\rho(h)$ is valid for an atmosphere in the state of thermal and mechanical equilibrium in a uniform gravitational field ($T = \text{const}$ and $g = \text{const}$ throughout the atmosphere). It is admissible to consider the strength of the gravitational field $g$ (the acceleration of free-fall) to be independent from the height $h$ when the characteristic height $H$ of the atmosphere is small compared to the planet's radius $R$: $H \ll R$. This condition is clearly fulfiled for the Earth's atmosphere and the atmospheres of other planets of the Solar System. The assumption concerning the thermal equilibrium of the atmosphere is a more serious restriction to applicability of (4.4). An improvement of this simplified model of the atmosphere is achieved if the characteristic height $H$ in (4.4) is considered as a function of the altitude $h$ because air temperature $T$ changes with altitude (see section 10.6 for details).

The exponential dependence expressed by (4.4) means a very rapid decrease of air density with altitude, and yet the density of air up to altitudes of 160 km is such that it does not allow satellites to orbit the Earth for a prolonged time. The higher the altitude of the orbit, the longer the satellite remains in orbit.

For the upper strata, this model of an equilibrium, exponential atmosphere is only approximate. At high altitudes the density of air depends strongly on the local temperature and on the conditions of illuminance by the Sun's radiation. The density at a given altitude changes considerably with the 24 h cycle, as though in daylight the upper atmosphere thickened, increasing its density under the Sun's rays. At the height of 350 km over the Earth's surface, the density in the daytime is 1.2 times greater than at night, and at the height of 500 km it is approximately 3 times greater.

In the program, the characteristic height of the atmosphere is a parameter which we can vary widely. For instance, we can give an exaggeratedly high value to $H$ in order to make the effect of air resistance much greater than it is in real situations. Such simulation experiments can help us to better understand the effect of air resistance in space flights around the Earth or other planets.

The resistance of air experienced by a satellite in the upper strata is proportional to the local density of the atmosphere, to the square of the velocity of the satellite and to its cross-sectional area. It depends also on the shape of the satellite. A detailed explanation of these properties of the air-drag force is given in section 10.6.

The acceleration produced by this force of atmospheric drag is inversely proportional to the mass of the satellite. Consequently, this perturbational acceleration can be expressed by:

$$\mathbf{a} = -C\rho(h)v\mathbf{v}, \tag{4.5}$$

in which the coefficient $C$ is proportional to the cross-sectional area of the satellite and inversely proportional to its mass. It has especially large values for light, hollow satellites like balloons inflated with a gas. Such satellites are subjected to air resistance (as well as to the radiant pressure of sunlight) in a greater degree than are massive compact bodies and so can be used for the experimental investigation of the upper atmosphere.

In describing the motion of a satellite, we refer it to the geocentric frame whose axes are fixed with respect to remote stars. This frame does not rotate in space, although its origin moves around the Sun together with the Earth in an almost circular orbit. The unperturbed geocentric (circular or elliptical) orbit of the satellite lies in a plane whose orientation is fixed in space relative to the stars and does not change in time. However, the surface of the Earth and the atmosphere move relative to the plane of the orbit due to the 24 h rotation of the Earth about its axis.

The acceleration $\mathbf{a}$ given by (4.5) is directed against the velocity $\mathbf{v}$ of the satellite *relative to the air*. The atmosphere is involved by the Earth in its axial rotation, so that $\mathbf{v}$ in (4.5) is actually the vectorial difference between the velocity of the satellite's orbital motion and the velocity of the atmosphere produced by the 24 h rotation of the Earth. Therefore the perturbational acceleration $\mathbf{a}$ of the satellite has a component perpendicular to the plane of its (non-equatorial) orbit. However, it can be shown that the change in orientation of the orbit caused by this component of the acceleration during the whole life of the satellite is negligible. Therefore in the simulation program it is assumed that $\mathbf{v}$ in (4.5) is the orbital velocity of the satellite.

The second parameter to be entered in simulating motion under air resistance (called here the air-drag coefficient) is the ratio of the acceleration produced by air resistance when the satellite moves through the atmosphere at $h \approx 0$ with a velocity equal to the circular velocity at $h \approx 0$, to the gravitational acceleration (acceleration of free-fall) near the surface of the Earth. This dimensionless parameter depends both on the density of the atmosphere near the surface and on the characteristics of the satellite (its mass and cross-sectional area). The program enables us to compare the influence of the atmosphere on satellites with different mass-to-cross-sectional-area ratios by simultaneously launching several satellites from the same spatial point and with the same initial velocity, but with different values of the air-drag coefficient.

The program also allows us to compare the actual motion of a satellite or a missile under the influence of air resistance, with the corresponding unperturbed motion governed only by gravity. At any instant during the simulation, we can cause the program to display the osculating ellipse along which the projectile would continue to move if air resistance were absent, starting from the current values of the projectile position and velocity (which are used as the initial conditions for the calculation of the subsequent unperturbed motion).

To implement this feature, at the beginning of the simulation we select the option 'Unperturbed curve' from the menu. When this item is checked, the program draws the osculating ellipse each time we click the button 'Pause'. When we resume the simulation after the ellipse is displayed, the difference between the actual motion under the air drag and the idealized unperturbed Keplerian motion is evident.

If we have chosen to launch several satellites (conditions for which are listed in a table), the program automatically chooses the scale that simultaneously displays all the (unperturbed) trajectories in the window. However, we can override the automatic scaling and customize the scale in order to make some specified orbit fit the window (other orbits at this scale may be partially cropped by the borders of the window). To do so, we switch off the option 'Autoscaling' in the panel 'Input', and select some orbit in the list. We next click the button 'Rescale'. The program chooses an optimal scale to display the selected orbit in the window.

Using the 'Input' panel, we can edit the list of chosen launches. To remove some orbit from the simulation, we select the corresponding line in the list and click the 'Remove' button. To add a new orbit, we enter the necessary parameters by typing the values into the corresponding boxes, or by dragging the slides on the scroll-bars. We next click the button 'Add'. A new line with the parameters that are currently displayed in the boxes is added to the list of launches.

When we make a sequential simulation of several launches, it is also possible to display the motion along the simulated orbit in a separate window. To open such a window with an increased image of the orbit, we check the menu item 'Zoom'. We can change position of this window on the screen and resize it for more comfortable observation.

## Questions and problems

1. (*) **Evolution of the extended elliptical orbit in the upper atmosphere.** Explain why the apogee of a strongly stretched elliptical orbit whose perigee is located in the rarified upper strata of the atmosphere descends gradually with every revolution of the satellite, so that the shape of the orbit tends gradually to a circle.
2. (*) **Evolution of the circular orbit in the upper atmosphere.** A satellite orbiting the Earth in a low circular orbit is continuously subjected to a very small air resistance caused by the rarefied upper strata of the atmosphere. After a large number of revolutions, the radius of the satellite's orbit (the distance from the center of the Earth) has diminished by 0.2% (that is, by a factor of 0.002). By what factor has the velocity of the satellite changed during this time? By what factor has the period of revolution changed? Assume that each individual loop of the satellite's trajectory can be considered as almost a closed circle.
3. (*) **Aerodynamical paradox of the satellite.** Explain why the retarding force of air resistance in the upper strata of the atmosphere causes an increase in the orbital velocity of the satellite. How can you reconcile this fact with the law of energy conservation?

# Chapter 5

# Active maneuvers in space orbits

Many interesting problems in space dynamics are associated with modifying the orbit of a satellite or a spacecraft in order to produce a particular trajectory for an intended space flight. The orbit can be modified by applying a brief impulse to the craft. In particular, the velocity of the craft can be changed by the thrust of a rocket engine that is so oriented and of such duration as to produce the desired result. The maneuver should be executed at a proper instant by the astronauts of the spacecraft or by a system of remote control.

When the engine is very powerful and operates for a very short time (so short that the spacecraft covers only a very small part of its orbit during the thrust), the change in the orbital velocity of the spacecraft is essentially instantaneous. In this simulation it is assumed that this change occurs instantly. After such a maneuver the spacecraft continues its passive orbital motion along a new orbit. The parameters that characterize the new orbit depend on the initial conditions implied by momentary values of the radius vector and the velocity vector of the spacecraft at the end of the applied impulse.

With the simulation program that deals with maneuvering a spacecraft, we can design a space flight beforehand and then carry out its simulation, playing the role either of a space pilot or of a distant operator using remote control. The necessary maneuvers of the spacecraft are executed automatically according to the program that we have entered, or we do them manually by instantly changing the velocity vector of the craft at appropriate moments in the course of the simulation.

## 5.1 How to operate the program

It is assumed that originally the spacecraft is docked at a permanent station that orbits the Earth (or some other planet) in a circle. Both the height of this orbit and the magnitude of the additional velocity of the spacecraft are to be entered beforehand. This additional velocity (sometimes called the *characteristic velocity*) is imparted by the rocket engine to the spacecraft after it is undocked.

doi:10.1088/978-0-750-31100-7ch5  

There are two ways, automatic and manual, to control the instant at which this maneuver takes place. For automatic control, we enter the time beforehand (during the design time) by typing the time into the box 'Time of maneuver' of the input panel, which we can open by choosing the corresponding item in the menu. The time zero corresponds to the instant at which the simulation starts. We can choose either natural units of time and velocity, namely, the period of the station in its circular orbit and the corresponding circular velocity, or the usual units (seconds and kilometers per second). When we have chosen the magnitude of the additional velocity and indicated the time for the maneuver, we click the button 'Add' in order to include the values in the list of ordered maneuvers. The direction of the additional velocity for this maneuver depends on which of the options ('Up', 'Down', 'Forward', or 'Backward') is chosen in the frame 'Direction'. This direction can be fixed either relative to the local horizon, or relative to the vector of instantaneous velocity of the spacecraft, depending on the choice of option in the frame 'Orientation'.

In this way we can order as many maneuvers as we like, adding in sequence the required values of the additional velocity and its direction, as well as the time the maneuvers start, to the list of ordered maneuvers. We can edit the list not only by adding new items but also by removing some items (to do so, we select an item and click the button 'Remove'), or by inserting additional items. To insert an item, we indicate the required additional velocity and the time for the maneuver in the corresponding boxes, and in the list we select the item before which we want to insert the new item. We then click the 'Insert' button.

When the list is ready, we click the 'Ok' button. The simulation program performs the ordered maneuvers if we choose the option 'Reproduce' in the frame 'Maneuvers' and click the 'Start' button.

We can avoid a preliminary calculation of the time at which each maneuver is to begin if we use the manual control. Watching the motion of the space station on the computer screen, we can click one of the four arrow command buttons in the upper panel of the window in which the motion is displayed. We thereby give a command to undock the space vehicle from the orbital station and to instantly add velocity to the vehicle by the thrust of its rockets. In the simulation, this additional velocity is imparted to the spacecraft just at the moment we click the corresponding command button. The orientation of the additional velocity $\Delta\mathbf{v}$ depends on which of the four arrow buttons we click.

Vector $\Delta\mathbf{v}$ of the additional velocity lies in the plane of the orbit and can have one of four orientations: up, down, forward, or backward. These directions of $\Delta\mathbf{v}$ are defined either with respect to the vector $\mathbf{v}$ of the instantaneous velocity of the spacecraft, or with respect to the local vertical line at the point of the maneuver, depending on the option chosen before the maneuver. The magnitude of the additional velocity is determined by the value which has been entered in the corresponding box of the panel 'Input' before the simulation.

During the subsequent simulation, we can compare the motion of the space vehicle with the motion of the orbital station, which remains in a circular orbit around the planet.

We can perform several orbital maneuvers during the space flight, each time using the manual control to pick the instant the rocket engine is to start and the direction of the additional impulse. However, the magnitude of the additional velocity for each of the maneuvers must be chosen beforehand and added to the list of ordered maneuvers in the panel 'Input'. We do this in the same way as we do for the automatic control, when we enter the height of the original circular orbit and other parameters of the simulation. For each consecutive maneuver, the magnitude of the additional velocity equals the next value in the list.

If we click one of the arrow command buttons while the program is executing the sequence of maneuvers ordered beforehand, we force the program to automatically switch to the mode of manual control. This means that the instants of subsequent maneuvers, as well as the directions of the additional velocity, indicated in the list of ordered maneuvers are ignored by the program. We determine these instants during the simulation manually by clicking on the arrow buttons at the appropriate times.

If manual control is to be used during the simulation, it is not necessary to indicate beforehand the instants and directions of the additional velocity for the sequence of proposed maneuvers. (If we have done so nonetheless, manual control overrides the values entered earlier.) We just input the magnitude of the additional velocity for a maneuver into the corresponding box and leave the box 'Time of the maneuver' empty. We next click the 'Add' (or 'Insert') button and the value of the additional velocity is included in the list of maneuvers with an indefinite value for the time. It is assumed that the time will be specified manually during the simulation.

During the simulation, when the list of ordered maneuvers is exhausted (it makes no difference whether the mode is automatic or manual control), we can nevertheless continue performing maneuvers manually. In this case the last value of the magnitude of the additional velocity in the list is used each time one of the arrow buttons is clicked.

Under manual control, the program stores in memory the sequence of maneuvers that are performed. All their characteristics, including the times and directions of the additional velocity, are included in the list. This sequence of maneuvers can be reproduced afterward by clicking on the 'Restart' button and choosing the option 'Reproduce' in the frame 'Maneuvers'. (To enable the option button 'Reproduce' during the simulation, we should click on the 'Pause' button.)

Before restarting the simulation, we can edit the list of maneuvers and make the necessary corrections to the parameters. We do this in the way described above by using the panel 'Input', which can be opened by choosing the corresponding item in the menu.

Working with the program, we can open an additional window in order to simultaneously display the motion of the space vehicle in another frame of reference. This additional frame is associated with the orbital station. More precisely, this frame is fixed to the rotating straight line joining the center of the planet with the orbital station, so that one of the coordinate axes is always directed along this line. Since this frame rotates uniformly around the planet together with the orbital station, it is a non-inertial frame of reference, in which the orbital station is at rest.

In other words, the relative motion of the space vehicle displayed in this window shows the motion as seen by the astronauts observing it from the orbital station.

In the unusual conditions of orbital flight, navigation is quite different from what we are used to on Earth, and our intuition fails us. The relative motion of the spacecraft observed in this additional window reveals many extraordinary features that are hard to reconcile with common sense and our everyday experience.

To open the window that displays the relative motion, we click the menu item 'Zoom', 'Relative Motion'. If we like, we can resize it and move around the screen in the usual way. If the scale chosen automatically by the program is not satisfactory, we can change it by clicking on the menu items 'Rescale', or 'Zoom in' and 'Zoom out'. The item 'Rescale' is useful when we have already simulated the motion but wish to repeat it with the same parameter values.

It is recommended that when starting this program, we look through the set of ready examples with which the program is provided. These examples illustrate various possible orbital maneuvers. To open the panel with the list of examples, we click the menu item 'Examples'. With the help of the menu 'Options' of the panel 'Examples', we can open either a set of basic examples, or an advanced set with more sophisticated topics. To run an example, we select it from the list and click the 'Ok' button. We next click the button 'Start' in the main panel. We do not need to enter data to run the simulations from these examples. Further on, we can create our own examples and store them as additional (customized) sets.

## 5.2 Space flights and orbital maneuvers

The aims of orbital maneuvers may be varied. For example, we may plan a transition of the vehicle into a higher circular orbit in order to remain in it for some time, eventually returning to the orbital station and soft docking to it. Or we may wish to design a transition of the landing module to a descending elliptical orbit that grazes the Earth's surface (the dense strata of the atmosphere) in order to return to the Earth from the initial circular orbit. We may want to launch from the orbital station an automatic space probe that will explore the surface of the planet from a low orbit, or, on the other hand, to send a probe far from the Earth to investigate interplanetary space. The orbit of the space probe must be designed to make possible its return to the station after the necessary investigation is over.

### 5.2.1 Designing a space flight

To plan such space flights, we must solve various problems related to the design of suitable transitional orbits. We must decide how many instant maneuvers are necessary to reach the goal. To make each transition of the space vehicle into a desired orbit, we must calculate beforehand the magnitude and direction of the required additional velocity (the *characteristic velocity*), as well as the time at which this velocity is to be imparted to the space vehicle. As a rule, the solution of the problem is not unique.

The complexity of the problem arises from the expectation that we choose an *optimal* maneuver from many possibilities. The problem of optimization may include various requirements and restrictions concerning admissible maneuvers. For example, there may be a requirement of minimal expenditure of rocket fuel, with an additional condition that possible errors in navigation and control (in particular, errors in determining the time for the maneuver and small errors in the direction of the additional velocity) do not cause inadmissible deviations of the actual orbit from the calculated one.

### 5.2.2 The way back from space to the Earth

As an example of active maneuvers of a spacecraft staying originally in a low circular orbit around a planet, let us consider the problem of transition of a landing module to a descending trajectory. For a safe return to the Earth, the landing module must enter the dense strata of the atmosphere at a very small angle with the horizon. A steep descend is dangerous because of the rapid heating of the spacecraft in the atmosphere. The thermal shield of the landing module must satisfy very stringent demands. For a manned spacecraft, large decelerations caused by the air drag at a steep descend are inadmissible mainly because of the dangerous increase in the pseudo-weight of the space travelers. All this means that the planned passive descending trajectory must just graze the upper atmosphere.

Next we shall consider and compare two possible ways to transfer the landing module into a suitable descending trajectory.

1. After the landing module is undocked from the orbital station, it is given an additional velocity directed opposite to the initial orbital velocity.
2. The additional velocity of the landing module is directed downward (along the local vertical line).

In all cases, any additional velocity transfers the space vehicle from the initial circular orbit to an elliptical orbit. One of the foci of the ellipse is located, in accordance with Kepler's first law, at the center of the Earth.

In the first case, brief operation of the rocket engine changes only the magnitude of the orbital velocity, preserving its direction. Therefore, at the point where the rocket engine operates (point $A$ in figure 5.1), the descending elliptical orbit has a common tangent with the original circular orbit. This point $A$ is the apogee of the

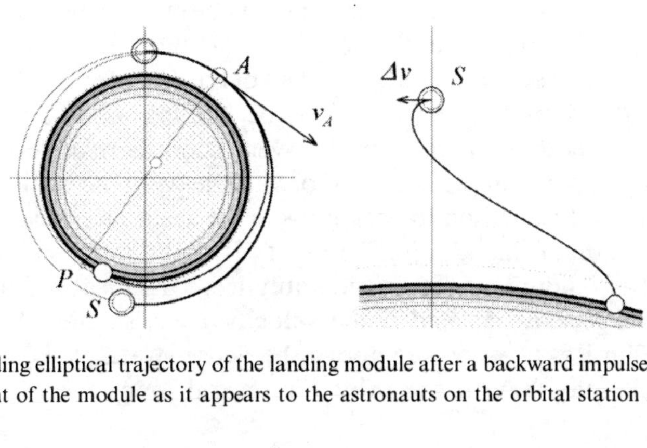


**Figure 5.1.** Descending elliptical trajectory of the landing module after a backward impulse is applied at point $A$ (left) and the descent of the module as it appears to the astronauts on the orbital station $S$ (right).

elliptical orbit. Its perigee is located at the opposite end $P$ of the major axis, that passes through $A$ and the center of the Earth. It is evident that precisely at this point $P$ the ellipse must just graze the surface of the planet (more exactly, the ellipse grazes the dense strata of the atmosphere). The landing module must enter the atmosphere near this point of the descending orbit.

The additional (characteristic) velocity $\Delta v$ necessary for the transition from the circular orbit to this elliptical trajectory can be calculated from the conservation laws of energy and angular momentum. Details of the calculation can be found in section 10.5. Here we give the resulting formula:

$$\Delta v = v_c\left(1 - \sqrt{\frac{2}{1 + r_0/R}}\right). \tag{5.1}$$

Here $v_c$ is the circular velocity of the space station, $r_0$ is the radius of this orbit and $R$ is the Earth's radius (more exactly, radius of the Earth together with the atmosphere).

In the case of a low circular orbit, whose height $h$ over the Earth is small compared to the Earth's radius ($h \ll R$), the exact equation (5.1) can be replaced by an approximate expression:

$$\Delta v \approx v_c \frac{h}{4R}. \tag{5.2}$$

For example, if the height $h$ of the circular orbit equals $0.2\ R \approx 1270$ km, the additional velocity $\Delta v$, according to (5.2), must be about 5% of the circular velocity. (The calculation on the basis of (5.1) with $r = R + h = 1.2\ R$ gives a more exact value of 4.65%).

This method of descending from a circular orbit (with the help of a backward impulse) requires the absolutely minimal amount of rocket fuel. However, it is also very sensitive to small variations in the value of the additional velocity. In the ideal situation, if the additional velocity has exactly the required value given by (5.1), the point of landing is near the perigee $P$ of the ellipse (see figure 5.1). During the descent, the landing module covers just one half of the ellipse (from $A$ to $P$) while the station covers a little less than half its circular orbit. At the moment of landing, the station is above and a little behind the module (point $S$ in figure 5.1).

The right panel of figure 5.1 shows the descent of the module in the frame of reference associated with the orbital station. At first the landing module actually moves in the direction of the additional velocity, but very soon its relative velocity reverses. Gradually descending, the module moves forward, leaving the station behind.

The sensitivity of this method to variations in the additional velocity means that if the actual magnitude of the additional velocity is slightly greater than the required value, the point of landing moves considerably from the idealized perigee (point $P$) toward the starting point $A$. And if the velocity $\Delta v$ is smaller than required, the perigee of the elliptical orbit occurs above the dense strata of the atmosphere, and the space vehicle may stay in the orbit for several loops more. Because there is

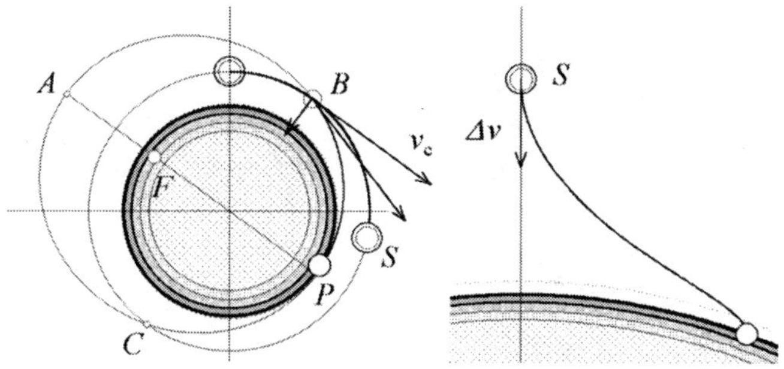


**Figure 5.2.** The descending elliptical trajectory of the landing module after being given a downward additional velocity at point *B* (left) and the descent of the module as it appears to the astronauts on the orbital station *S* (right).

considerable air resistance near the perigee, the apogee gradually descends after each revolution. The orbit approaches a low circle, evolving as described in the section 4.2 'Evolution of an Orbit in the Atmosphere'. Eventually the space vehicle enters the dense atmosphere and lands. However, it is almost impossible to predict when and where this landing will occur.

If the additional velocity imparted to the space vehicle at point $B$ of the initial circular orbit (figure 5.2) is directed *radially* (transverse to the orbital velocity), both the magnitude and direction of the velocity change. Therefore, the new elliptical orbit *intersects* the original circular one at this point $B$. For a soft landing, the new elliptical trajectory of the descent must also graze the Earth (the upper atmosphere) at the perigee $P$ of the ellipse. Using the laws of energy and angular momentum conservation (see section 10.5) and requiring that the perigee distance $r_P$ be equal to the Earth's radius $R$, we find that the necessary additional velocity $\Delta v$ for this method of landing is given by

$$\Delta v = v_c \frac{h}{R}. \tag{5.3}$$

Here $v_c$ is the circular velocity for the original orbit, $h$ is its height above the surface (above the atmosphere) and $R$ is the Earth's radius (including the atmosphere). Comparing this expression with (5.2), we see that for this method of transition to the landing trajectory, the required additional velocity is approximately four times greater than that for the first method. For example, it must equal 20% of the circular velocity if the height $h$ of the circular orbit is $0.2\,R$. The angular distance between the starting point $B$ and the landing point for this method equals 90° (a quarter of the revolution), in contrast to the first method, for which the angular distance between the point of transition from the circular orbit to the descending trajectory and the landing point is twice as large (half a revolution).

Figure 5.2 also shows the position $S$ of the orbital station at the moment of landing. We can see that the station is above and some distance behind the landing module since for the moment of landing the station has not completed a quarter of its revolution beyond the initial point $B$.

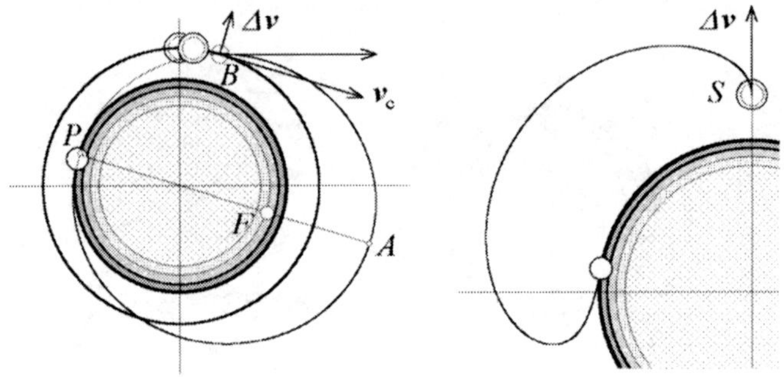


**Figure 5.3.** Elliptical trajectory of the landing module after acquiring an upward additional velocity at point *B* (left) and the trajectory of the module as it appears to the astronauts on the orbital station *S* (right).

The right panel of figure 5.2 shows the landing trajectory in the frame of reference associated with the orbital station. At first the astronauts on the station see that the landing module really moves downward, in the direction of the additional velocity imparted by the on-board rocket engine. However, soon the trajectory bends forward, in the direction of the orbital motion of the station. The landing module on its way toward the ground moves forward, leaving the station in its orbital motion far behind.

Strange as it may seem, we can transfer the space vehicle to a landing trajectory by a transverse impulse directed *vertically upward* as well as downward (figure 5.3). In this case, starting from the point *B* of transition to the elliptical orbit, the landing module first rises higher above the Earth. Only after it passes through the apogee *A* of the orbit does it begin to descend toward point *P* (the perigee of the orbit), at which it enters the atmosphere. The angular distance between the starting and the landing points (*B* and *P*, respectively) in this case equals 270°, that is, three quarters of a revolution. During this time, the orbital station covers almost a whole revolution and at the moment the vehicle lands it is far beyond the landing point.

The trajectory of the landing module as seen by the astronauts in the orbital station is shown in the right panel of figure 5.3 (for the case of $h = 0.3\ R$). The module first moves upward, in the direction of the additional velocity, but soon turns backward. Its relative motion becomes retrograde, and the landing module lags behind the station. After circling more than a quarter of the globe in the retrograde direction, the module's motion reverses direction. The module then descends, approaching the Earth's surface tangentially.

For an elliptical orbit that is to graze the Earth, the magnitude of the additional velocity must be the same for both the downward and upward directions of the impulse. We can easily see this point either from the laws of the conservation of energy and angular momentum (the corresponding equations are the same for both cases), or from considerations based on the symmetry between the two cases: for if the goal is to land the module at some point *P* of the Earth's surface (figure 5.2), we must make a transition from the initial circular orbit to an elliptical orbit for which point *P* is the perigee. The orbits intersect at two points *B* and *C*. The transition is possible either at *B* using a downward impulse, or at a symmetrical point *C* using an upward impulse of equal magnitude.

## Questions and problems

1. **Landing with minimal fuel consumption.** Explain why the most economical way of transiting from a circular orbit to the landing trajectory is provided by a backward impulse of the additional velocity.
2. **Additional impulse required for the maneuver.** Using the laws of conservation of energy and angular momentum (Kepler's second law), calculate the additional backward velocity $\Delta v$ required to transit from a circular orbit to the landing trajectory. Express your answer in terms of natural quantities (for the problem under consideration), namely, in terms of the circular velocity $v_c$ for the original orbit, radius $r$ of the orbit and radius $R$ of the planet.
3. **Descent from a low circular orbit.** For the case of a low circular orbit ($h = r - R \ll R$), simplify the expression for $\Delta v$ obtained above. In the approximate formula hold terms linear and quadratic in the small parameter $h/R$. Verify the approximate formula in the simulation experiment.
4. **Landing on a planet with an atmosphere.** Investigate experimentally the influence of atmospheric drag experienced by the landing module on the position of landing point. Try varying the height of the atmosphere and the air-drag coefficient that characterizes the landing module.
5. **Trajectory of the landing module relative the orbital station.** Explain the shape of the trajectory of the landing module in the reference frame associated with the orbital station. This trajectory is displayed in the additional window 'Relative motion' (see right panels of figures 5.1 and 5.2).
6. **Downward additional impulse for the landing module.** On the basis of the laws of conservation of energy and angular momentum, calculate the additional velocity $\Delta v$ that must be imparted to the landing module in the downward vertical direction in order to transfer the module to the landing trajectory. Express the answer in natural units—the circular velocity $v_c$ for the initial orbit and the ratio $h/R$ of the height of the orbit to the Earth's radius.
7. **Landing after an upward additional impulse.** Explain why the additional velocity imparted to the landing module in the upward vertical direction must have the same magnitude $\Delta v$ as in the preceding case in order to obtain a trajectory that also grazes the surface of the Earth. Explain the shape of the trajectory observed in this case by the astronauts from the orbital station (right panel of figure 5.3).
8. **Descent along a vertical line**. What additional velocity should be imparted to the space probe undocked and launched from the orbital station that stays in a circular orbit in order for the probe to descend to the ground along a vertical line? What will be the trajectory of the probe as seen by the astronauts watching its motion from the orbital station?

## 5.3 Relative motion of bodies in space orbits

Let two satellites orbit the Earth. We know that their passive orbital motion obeys Kepler's laws. But how does one of them move relative to the other? This relative motion is important, say, for the spacecraft docking in orbit. If two satellites are brought together but have a (small) non-zero relative velocity, they will drift apart non-rectilinearly. In the unusual conditions of orbital flight, navigation is quite different from what we are used to here on the Earth and our intuition fails us. The study of the relative motion of the spacecraft reveals many extraordinary features that are hard to reconcile with common sense and our everyday experience.

### 5.3.1 The motion of a small body ejected from the orbital station

In this section we discuss the problem of the passive relative motion of orbiting bodies on the specific example of the free motion of any body that is ejected from an orbital station that stays in a circular orbit. It is essential for this problem that the initial velocity of the body relative to the station be small compared to the orbital velocity of the station.

The simulation program that displays the relative motion allows us to get an impression of such motion. For example, what does the motion of a body thrown vertically down toward the Earth look like for the astronauts on the orbital station?

Our thinking about this problem may pass through several stages.

The first stage is likely to be governed by jumping too quickly to a conclusion: relying on our everyday experience, we find nothing strange if the body thrown toward the Earth simply falls rapidly toward the ground.

The second stage begins with a reflection that the orbital station travels over the Earth at a great speed—more than seven kilometers per second! What is the initial velocity of the body that is thrown from the moving station by an astronaut? A healthy man can throw a small stone with a speed of about 10–20 m s$^{-1}$.

Considering the motion of the body relative to the Earth, we should add vectorially the velocity of the orbital station to the velocity of the body with respect to the station. We see that the resulting velocity of the body differs only negligibly in magnitude and direction from the velocity of the station. This means that after being thrown, the body simply transfers to another orbit around the Earth and this new orbit is almost indistinguishable from the orbit of the station. Does this conclusion agree with the first conclusion, that the body rapidly recedes from the station toward the Earth?

Next our speculations pass over to the third, exploratory stage. We recall reasonably that the question is concerned not with the motion of the body relative to the Earth, but rather with the motion as it is seen by the astronauts on the station. In other words,

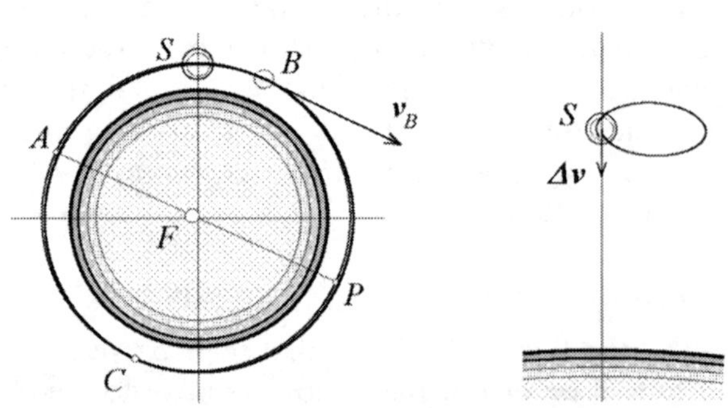


**Figure 5.4.** The motion of the body thrown from the orbital station down toward the Earth, as this motion is seen from the Earth (left) and from the orbital station (right).

we should investigate the motion of the body in the frame of reference associated with the orbital station. The simulation program allows us to observe this relative motion on the computer screen (figure 5.4).

In the right panel of the figure we see that relative to the station the body indeed moves at first downward, in the direction of the additional velocity. However, soon the trajectory turns forward, then upward and backward, and finally the body returns to the station from the opposite side (from above), tracing an almost closed trajectory!

To explain the physical reasons of the relative motion of the orbiting bodies, we should consider first the motion of the body and the station relative to the Earth. This motion is shown in the left panel of figure 5.4. Because of the additional velocity directed toward the center of the Earth which the body received at point $B$, the ejected body transfers to an elliptical orbit of very small eccentricity. One of the foci is located at the center of the Earth and the second at a point $F$ that is very close to the center. This ellipse almost merges with the circular orbit of the station. We can barely see that only near perigee $P$ (figure 5.4) is the ellipse slightly inside the circle. Near apogee $A$ it is slightly outside the circle. With great precision we can treat this ellipse as a circle of the same radius, whose center is displaced from the center of the Earth toward $F$ through the distance half way to $F$.

Since the diameter of the circle and the major axis of the ellipse are almost equal, the periods of revolution for the station and for the body, according to Kepler's third law, are nearly equal. Both trajectories intersect one another at the initial point $B$ and at the opposite point $C$. At this point, the body is again at the same height as is the station. The station comes to this point $C$ after exactly half the period of its uniform rotation around the Earth.

However, the motion of the body along its elliptical orbit is slightly non-uniform: in accordance with Kepler's second law, the body comes to point $C$ a bit earlier than the station, since at the middle of this half of its orbit the body passes through the perigee $P$, where its speed is slightly greater than that of the station. As a result, at the common point $C$ of both orbits the body is in front of the station. At this moment the distance between the body and the station reaches a maximum.

During the second half of the revolution, the body passes through the apogee $A$ of its elliptical orbit. Because its speed decreases slightly in this part of the orbit, the body comes to the common initial position almost simultaneously with the station, approaching it from above. Hence, the motion of the body relative to the station occurs along an almost closed trajectory. One cycle of this motion is completed during a period of revolution of the station in its circular orbit.

Are the astronauts on the orbital station really able to observe this periodic motion of the body? Here we get on to the fourth stage of our investigation—to calculations. Let us next evaluate the linear dimensions of the almost closed loop covered by the body in its motion relative to the station (see the right panel of figure 5.4).

As we already mentioned, the geocentric elliptical orbit of the body can be approximated by a circle whose center is displaced from the center of the Earth along the major axis $PA$ through the half-distance between the foci. We can easily evaluate this displacement $\Delta x$, taking into account that the vectors of velocities for both orbits at the point of their intersection $B$ are perpendicular to the corresponding radii. (Neither these vectors nor the radii are shown in figure 5.4, since the angle $\Delta v/v_c$ between them is so small that they merge in the image.) Consequently, the displacement of the center equals the radius $r_0$ of the circle times the angle $\Delta v/v_c$ between the velocities of the body and the station at point $B$ (or at point $C$): $\Delta x = r_0(\Delta v/v_c)$.

The divergence between the two orbits is greatest near points $P$ and $A$ (figure 5.4) and just equals the displacement $\Delta x$ of the center calculated above. Hence, the lowest point $P$ of the elliptical orbit is lower than the circular orbit of the station by the distance $\Delta x$, and the highest point $A$ is higher by the same distance $\Delta x$. Consequently, the vertical size of the loop in the right panel of figure 5.4 (the minor diameter of the relative trajectory) equals $2\Delta x = 2r_0(\Delta v/v_c)$.

### 5.3.2 Numerical estimates

We are now ready to make a numerical estimate of the typical size of the trajectory that the body traces relative to the orbital station. Let, for instance, the height of the circular orbit of the station be $h = 0.1\ R \approx 640$ km (radius of the orbit $r_0 \approx 7000$ km, period of revolution $T \approx 98$ min) and the relative initial velocity of the body be $\Delta v = 15\ \text{m s}^{-1}$, that equals approximately 0.2% of the orbital velocity $v_c = 7.5\ \text{km s}^{-1}$. In this case the minor diameter of the relative trajectory equals $2\Delta x = 2r_0(\Delta v/v_c) \approx 28$ km. This value gives an idea of the actual sizes of the loop in figure 5.4.

It is unlikely that the astronauts would be able to see a small body at a distance of more than a kilometer. Therefore they can watch it only during the first part of its nearly closed trajectory. Most likely, they lose sight long before the deviation from its rectilinear downward motion becomes apparent. Thus, until the astronauts lose sight of the body, it simply falls down toward the ground, moving in the direction of the relative initial velocity imparted to the body by the astronaut!

It is also interesting to calculate the size of the relative trajectory in the horizontal direction. As mentioned above, the major diameter of the loop equals the lag of the station behind the body near the common point $C$ (see the left panel of figure 5.4) of their orbits. With the help of Kepler's second law, we can calculate the difference $\Delta T$ of the time intervals required for the station and for the body to cover the half-orbit between common points $B$ and $C$.

For the station, the area $A$ swept out by the radius-vector in its rotation from $B$ to $C$ during the time $T/2$ is exactly half a circle: $A = \pi r_0^2/2$. For the body, whose nearly circular orbit is displaced from the Earth's center through $\Delta x$, the corresponding area is smaller by $\Delta A = 2r_0\Delta x = 2r_0^2\Delta v/v_c$. Hence $\Delta T/(T/2) = \Delta A/A = 4(\Delta v/v_c)/\pi$, and $\Delta T = 4(r_0/v_c)(\Delta v/v_c)$ (since $T = 2\pi r_0/v_c$). Thus, for the maximal lag $l_{max}$ of the station behind the body at point $C$ we obtain:

$$l_{max} = v_c\Delta T = 4r_0\frac{\Delta v}{v_c}. \tag{5.4}$$

Consequently, the major diameter of the almost closed trajectory of the body in its motion relative to the station is twice its minor diameter. For the values that we used above (the height of the station $h = 0.1\ R \approx 640$ km, the relative initial velocity of the body $\Delta v = 15\,\mathrm{m\,s^{-1}}$, that equals approximately 0.2% of the orbital velocity $v_c = 7.5\,\mathrm{km\,s^{-1}}$), the maximal distance $l_{max}$ of the body from the station is approximately 56 km.

Section 11.7 contains a detailed derivation of the approximate differential equations that describe analytically the relative motion of a body in the vicinity of the orbital station. For the given initial conditions (a small relative velocity directed downward), the particular solution of the equations predicts the relative motion to be approximately along an elliptical loop stretched horizontally. The semiminor axis $l$ of the ellipse equals the radius of the circular orbit $r_0$ times the ratio of the additional velocity $\Delta v$ to the orbital velocity $v_c$, namely $l = r_0(\Delta v/v_c)$. The semimajor axis of the ellipse is twice as long. This result is exactly consistent with the estimate of the vertical size of the relative orbit considered above, as well as with the estimate of maximal distance $l_{max}$ between the body and the station, given by (5.4).

### 5.3.3 The secular component of the relative motion

The character of the relative motion of the body thrown from the orbital station is quite different if the relative initial velocity has a component parallel to the orbital velocity of the station. The trajectory of the relative motion is no longer a closed curve even for very small values of the initial relative velocity. The body does not return to the station. In the relative motion, along with the periodic components, there is a non-periodic *secular term,* responsible for the regular, steady receding of the body from the station.

Figure 5.5 illustrates the trajectory of the relative motion of the body that is thrown forward from the orbital station. At first the body actually moves forward, but gradually it deviates upward and soon its motion relative to the station becomes retrograde. Next the body descends and the trajectory makes a loop. The body periodically returns to the same height as the station, but with each return lags more and more behind the station.

The general character of this relative motion is easily explained with the help of the geocentric frame of reference. The new orbit of the body is an ellipse grazing the circular orbit of the station only at the initial point $P$, which is the perigee of the elliptical orbit. Passing through the apogee $A$, the body is higher than the station.

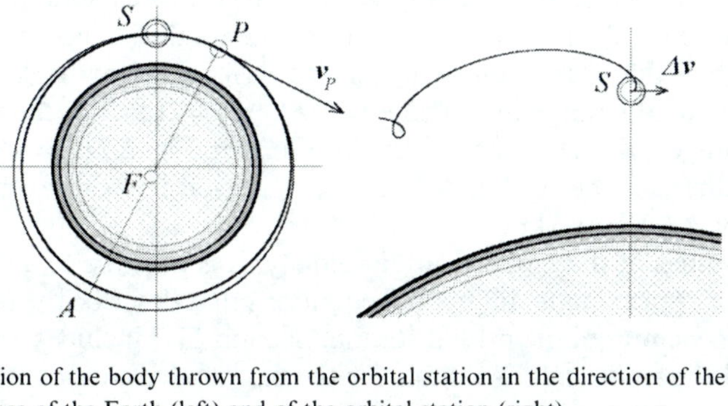


**Figure 5.5.** The motion of the body thrown from the orbital station in the direction of the orbital motion, in the frames of reference of the Earth (left) and of the orbital station (right).

Approaching the perigee $P$, the body descends to its initial altitude. But the period of revolution along the ellipse is greater than the period of the station. Therefore after a revolution the body arrives at the only common point $P$ of the two orbits later than does the station. This lag increases with each revolution. If the initial velocity of the body is such that the ratio of its period to that of the station is rational, the accumulated lag sooner or later becomes equal to the length of the whole orbit (or to several lengths of the orbit) and the body periodically meets with the station.

Another way of understanding the difference in character of the motion in the two cases (the additional velocity either along or perpendicularly to the orbital velocity) comes from the recalling that, for an inverse-square force, viewed from the reference frame associated with the source of the force, all orbits with negative energies are closed, and also that the period $T$ is uniquely related to the semimajor axis $a$ and hence to the total energy $E$—specifically, $T \sim a^{3/2} \sim (-E)^{-3/2}$ (see section 11.3 for details). No matter what the direction and magnitude of the velocity imparted to the ejected body, it will periodically return to and pass through the point in space from which it was thrown. But how will its period compare with that of the station? Throwing the object obviously has no effect on its initial potential energy. If it is thrown in a direction perpendicular to the orbital velocity of the station, then to the first order in $\Delta v/v_c$ its speed with respect to the geocentric frame will be unchanged, so its total energy (and likewise its period) will remain nearly identical to that of the station. That is, the time at which it returns to the spatial point from which it was thrown will be almost the same as the time of return of the station itself. If, however, its imparted velocity is directed along that of the station (whether it is forward or backward), its speed, and hence its total energy and period, will suffer a first-order change.

Indeed, for a circular orbit in the inverse-square central field, the (negative) total energy equals in magnitude the kinetic energy ($E = -E_{\text{kin}}$) and hence is proportional to the square of velocity: $-E \sim v^2$. Throwing the body in a direction orthogonal to the station's velocity will produce a second-order fractional change in the energy $\Delta E/(-E) = (\Delta v/v_c)^2$. Since $T \sim (-E)^{-3/2}$, this will result in the following fractional change in the period: $\Delta T/T = (3/2)(\Delta v/v_c)^2$. For the numbers used above ($\Delta v = 15\ \text{m s}^{-1}$, $v_c = 7.5\ \text{km s}^{-1}$), we find $\Delta T/T = 6 \cdot 10^{-6}$, which means that when the body will return to the starting point after a period, the station will be ahead of it in the orbit through a distance of about $v_c \Delta T \approx 265$ m. (We recall that in this motion, after a half-period $T/2$, the body occurs at a maximal distance of 56 km from the station.) On the other hand, throwing the body forward or backward produces a first-order fractional change in the energy $\Delta E/(-E) = 2(\Delta v/v_c)$, thus resulting in the fractional change of the same order in the period: $\Delta T/T = 3(\Delta v/v_c) = 6 \cdot 10^{-3}$. Therefore in this case the station occurs after a period in front of the body through the distance of $v_c \Delta T \approx 265$ km.

For small values of the initial relative velocity, it is possible to analyze the motion with the help of approximate differential equations that describe the relative motion of the body in the vicinity of the orbital station. Section 11.7 includes a derivation of the equations and their particular solution for the case under consideration.

## 5.4 Space probes and relative motion

As another example of a problem in which examining the relative motion of orbiting bodies is essential, we consider a space probe—an automatic or manned module with scientific instruments that is launched from a station circularly orbiting the Earth or some other planet. The module is to approach the surface of the planet in order to explore it from a low altitude. Then the module, with the scientific information it has collected, is to return to the orbital station and gently dock to it. Or, as another assignment, the space probe is to investigate far off regions of interplanetary space. In either case, its orbit must be chosen so that its rendezvous with the station is possible after it has completed its tasks.

Are such orbits possible? If so, how can they be realized?

After the space probe is undocked from the station and has almost instantly received an additional velocity with the help of the rocket engine, its further passive motion around the planet occurs along a new elliptical orbit. What requirements must this new orbit satisfy? If the probe is to investigate the surface of the planet, its orbit must approach the planet as closely as possible. Consequently, the orbit must have a low perigee (pericenter, for a planet in general), but must not intersect the surface (more precisely, the atmosphere) of the planet. Moreover, the period of revolution along such an elliptical orbit must be related to the period of revolution of the orbital station along its circular orbit in such a way that the probe and station periodically meet one another. Such a rendezvous can occur only at a common point of their orbits if their periods of revolution are in the ratio of integers, preferably small. For example, if the period of revolution of the probe is 2/3 the period of the station, the station completes two revolutions while the probe completes three. Thus the two meet at the common point of their orbits every two revolutions of the station after the departure of the probe.

### 5.4.1 Space probes in inner orbits

After the probe is undocked from the station, it moves along almost the same circular orbit and with the same velocity as does the station. In order to launch the probe into a required elliptical orbit, we should impart to it some additional velocity by means of an on-board rocket engine. From the point of view of rocket fuel expenditure, the most economical method of transition to a suitable orbit consists of imparting to the space probe an additional velocity tangent to its circular orbit. If the additional velocity is directed opposite to the orbital velocity of the station, we get an inner elliptical orbit that grazes the circular orbit of the station only at the orbital position at which the rocket engine thrusts the probe backward.

Let us consider the family of possible inner orbits.

The space probe encounters the orbital station each time the station completes a revolution if the period $T$ of the probe equals $T_0/n$, where $T_0$ is the period of the station, and $n$ is an integer. However, there is actually only one such possibility, namely, $n = 2$. Elliptical orbits with periods that equal $T_0/3$, $T_0/4$, $T_0/5$, ... do not exist. The reason is that the shortest possible period of revolution corresponds to the degenerate elliptical orbit with the minor axis of zero length (a straight-line ellipse

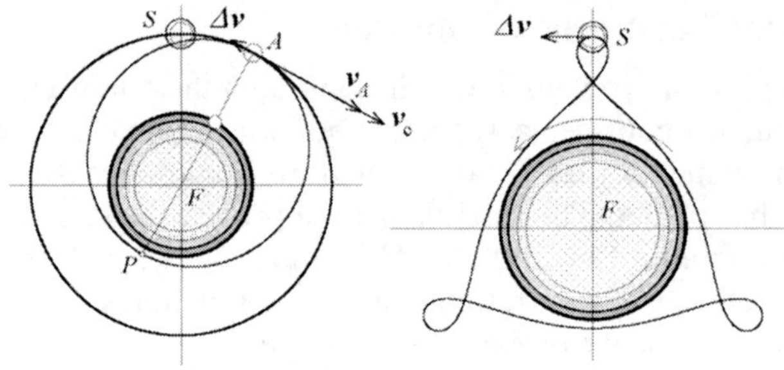


**Figure 5.6.** Elliptical orbit of the space probe with the period 2/3 $T_0$ (left) and the trajectory of the space probe in the reference frame associated with the orbital station (right).

with foci at the center of the planet and at the initial point, and with a major axis equal to the radius of the circular orbit). According to Kepler's third law, this minimal period equals $(1/2)^{3/2}T_0 \approx 0.35\ T_0$, a value greater than $T_0/3$.

For the elliptical orbit with the period $T = T_0/2$, the perigee distance $r_P$ equals $0.26\,r_A$, where $r_A$ is the apogee distance that equals the radius $r_0$ of the circular orbit of the station. Hence, the orbit can be realized only if the radius of the circular orbit is at least four times the radius of the planet. The characteristic velocity $\Delta v$ needed to transfer the space probe to this orbit from the initial circular orbit equals $0.36\,v_c$, that is, 36% of the circular velocity $v_c$. The formulas necessary for such calculations can be easily obtained on the basis of Kepler's laws and the law of energy conservation. The derivation of the formulas is given in section 10.5. The elliptical orbit of the space probe whose period equals 2/3 of the station period is shown in the left panel of figure 5.6. In this case a backward characteristic velocity $\Delta v$ of approximately $0.17\,v_c$ is required. At point $A$ the space probe is undocked from the station and the additional velocity $\Delta v$ is imparted to it by an on-board rocket engine. At the perigee $P$ of the elliptical orbit the distance $r_P$ from the center of the planet equals approximately $0.53\,r_0$. Hence, the orbit is ideal for a space probe if the circular orbit of the station has a radius approximately twice the radius of the planet.

The right panel of figure 5.6 shows the trajectory of this space probe in the rotating reference frame associated with the orbital station (more exactly, with the straight line joining the station and the center of the planet). For a while after the launch, the probe retrogrades in this frame in the direction of the additional velocity $\Delta v$. However, soon its trajectory turns first toward the planet and then forward—the probe overtakes the station in its orbital motion. As a whole, the trajectory of the probe bends around the planet in the same sense as the orbit of the station, in spite of the opposite direction of the initial velocity. Near the apexes of the loops of the trajectory the motion becomes retrograde. (These apexes correspond to the instants at which the space probe passes through the apogee of its geocentric orbit.) Moving along this closed trajectory, the probe approaches the surface of the planet three times. (At these instants the probe passes through the perigee of its geocentric orbit.)

To dock the space probe to the station after the voyage, we quench the remaining relative velocity (to equalize the geocentric velocities of the probe and the orbital station). This can be performed by the same on-board rocket engine. The required

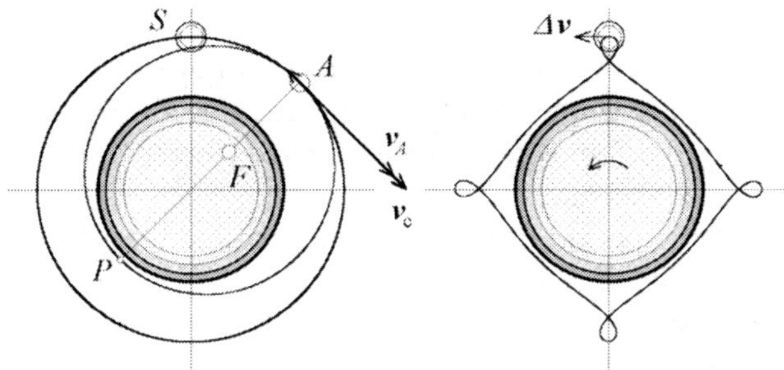


**Figure 5.7.** The elliptical orbit of the space probe with the period 3/4 $T_0$ (left) and the trajectory of the space probe in the reference frame associated with the orbital station (right).

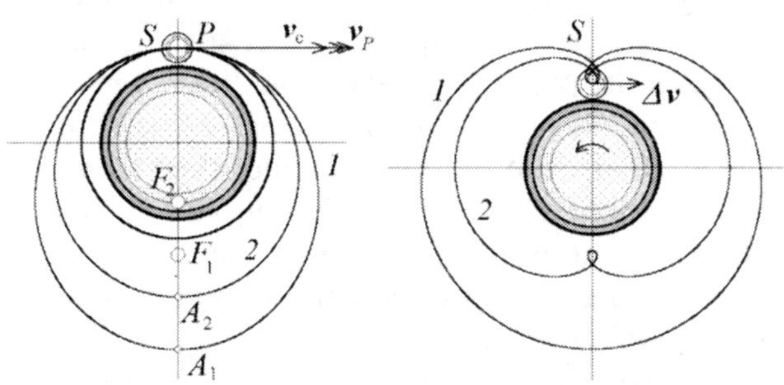


**Figure 5.8.** The elliptical orbits of the space probes with the periods 2 $T_0$ and 3/2 $T_0$ (left) and the corresponding trajectories in the reference frame associated with the orbital station (right).

additional impulse (the characteristic velocity of the maneuver) is just of the same magnitude as at the launch of the probe, but in the opposite direction: if at the launch the impulse is directed against the orbital velocity of the station, now at docking it is directed forward.

Figure 5.7 illustrates the motion of a space probe with the period of revolution $T = 3/4\ T_0$. In this case the probe meets the station at the initial point *A* after four revolutions around the planet. The station completes three revolutions during this time. The trajectory of motion of the probe relative to the orbital station has four loops that correspond to the instants at which the probe passes through the apogee of its orbit.

### 5.4.2 Space probes in outer orbits

If the additional velocity that is imparted to the space probe after its undocking from the station is directed forward, the elliptical orbit of the probe envelopes the circular orbit of the station, grazing it at the initial point. Figure 5.8 shows such outer elliptical orbits with the periods 2 $T_0$ and 3/2 $T_0$ (orbits *1* and *2*, respectively). The right panel of the figure shows the relative trajectories of the probe for these cases. At first the probe moves relative to the station in the direction of the initial velocity, but very soon its trajectory turns upward and then backward and the motion becomes retrograde—the probe lags behind the station. In this frame, the trajectory of the space probe bends around the planet in the sense opposite to the orbital motion of the station.

The trajectory with the period $T = 2\ T_0$ requires the additional velocity $\Delta v = 0.17\ v_c$, directed forward. The apogee distance $r_A$ (the maximal distance from the center of the planet) equals $2.17\ r_0$ ($r_0$ is the radius of the circular orbit). The closed orbit of the relative motion (curve *1* in figure 5.8) is covered during $2\ T_0$, that is, during two periods of revolution of the station.

For the orbit with the period $T = 3/2\ T_0$, the additional velocity is approximately $0.11\ v_c$, and the apogee distance is $1.62\ r_0$. The closed orbit of relative motion (curve *2* in figure 5.8) has two small loops, corresponding to the instants at which the probe passes through the perigee of its geocentric elliptical orbit. The whole closed path of the relative motion corresponds to two revolutions of the probe along the geocentric elliptical orbit, covered during three periods of revolution of the station.

In order to investigate both the surface of the planet and remote regions of interplanetary space using the same space probe, we can use an elliptical orbit obtained by imparting to the probe a transverse additional impulse. An example of such an orbit with the period of revolution $T = 3/2\ T_0$, is shown in figure 5.9. At point $B$ of the initial circular orbit the probe is undocked from the station, and the on-board rocket engine imparts a downward additional velocity $\Delta v$. The required magnitude of $\Delta v$ (the characteristic velocity of the maneuver) can be calculated on the basis of Kepler's laws and the law of conservation of energy. The corresponding calculation is found in section 10.5.

To obtain the orbit with the period $T = 3/2\ T_0$, a rather large characteristic velocity of $0.487\ v_c$ is necessary. (This value is several times larger than the tangential additional velocity of $0.11\ v_c$ needed for the elliptical orbit of the same period and the same major axis.) During three revolutions of the station, the space probe makes two revolutions along its elliptical orbit and they meet at the initial point $B$. To dock the space probe to the station softly, another additional impulse from the rocket engine is required. To equalize the orbital velocities, an additional velocity of the same magnitude $\Delta v$ as at the launch must be imparted to the space probe, but now it should be directed radially upward. The relative motion of the space probe in this case is shown in the right panel of figure 5.9. In the reference frame associated with the station, the space probe covers its convoluted closed path during three revolutions of the station around the planet.

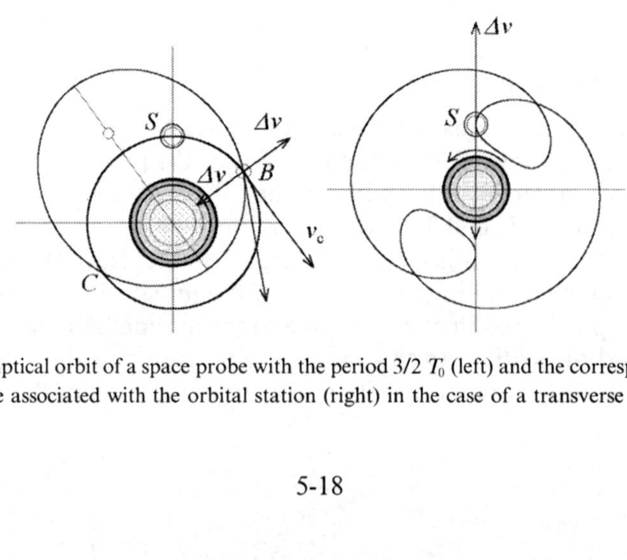


**Figure 5.9.** The elliptical orbit of a space probe with the period 3/2 $T_0$ (left) and the corresponding trajectory in the reference frame associated with the orbital station (right) in the case of a transverse additional impulse.

Let us consider one more example of space maneuvers. Imagine we need to launch a space vehicle from the orbital station into the same circular orbit as that of the station, but there is to be an angular distance of 180° between the vehicle and the station. In other words, they are to orbit in the same circle but at opposite ends of its diameter. How can this be done?

The task cannot be solved by a single maneuver. The on-board rocket engine must be used at least twice. With two impulses we can transfer the space vehicle to the opposite point of the circular orbit using an intermediate elliptical orbit with the period of revolution, say, $3/2\ T_0$ or $3/4\ T_0$. In the first case, after undocking from the station, an additional velocity $\Delta v = 0.11\ v_c$ is imparted to the space vehicle in the direction of the orbital motion. During one revolution of the space vehicle along its elliptical orbits (curve *2* in figure 5.8), the station covers exactly one and a half of its circular orbits. That is, the space vehicle reaches the common point $P$ of the two orbits (circular and elliptical) just at the moment when the station is at the diametrically opposite point of the circular orbit.

In the relative motion, shown in the right panel of figure 5.8, the space vehicle has covered one half of its closed path *2*. At this moment, we quench the excess of velocity of the space vehicle over the value $v_c$ by a second jet impulse and the vehicle moves along the same circular orbit as the station but at the opposite side of the orbit. In the window of the simulation program that displays the relative motion, the space vehicle is stationary at the antipodal point. Clearly the second jet impulse must be of the same magnitude as the first one but opposite to the orbital velocity.

## Questions and problems

1. **Relative motion after a vertical initial velocity.** An astronaut on an orbital station throws a small stone vertically upward with a velocity of $10\ \mathrm{m\ s^{-1}}$ relative to the station that moves in a circular orbit at an altitude of $0.1\ R$ ($R = 6400$ km is the Earth's radius.) What is the trajectory traced by the stone relative to the station? Calculate the maximal distance of the stone from the orbital station.
2. **Relative motion after a horizontal initial velocity.** An astronaut throws a stone with an initial velocity of $10\ \mathrm{m\ s^{-1}}$ directed perpendicularly to the plane of the circular orbit of the station that moves in a circular orbit at an altitude of $0.1\ R$ ($R$ is the Earth's radius.) What is the trajectory traced by the stone relative to the station? Calculate the maximal distance of the stone from the orbital station.
3. (**) **To reach and catch up, we should brake.** An orbital station moves around the Earth in a circular orbit. A spacecraft is launched to dock to the station, but because of a delay at the launch, the craft moves into the same circular orbit some distance $L$ behind the station. This distance is small compared to the radius $r$ of the orbit ($L \ll r$). In order that the spacecraft reach the station after one revolution along the orbit, an additional rocket impulse is required.
   (a) What should be the direction of the impulse? Calculate the additional velocity $\Delta v$ that must be imparted to the craft. Express it in terms of the distance $L$ and the period $T$ of revolution of the station. Also express $\Delta v$ in terms of the orbital velocity of the station $v_c$ and the ratio $L/r$.

(b) When the craft reaches the station, one more rocket impulse is required to equalize their velocities for soft docking. What is the additional velocity required for this maneuver?
(c) Repeat parts (a) and (b) for the case in which the spacecraft is to approach and dock to the station after two revolutions of their orbit.
(d) Repeat parts (a) and (b) for the case in which the spacecraft is in front of the station at a distance of $L$ ($L \ll r$).

## 5.5 Rendezvous in space and interplanetary flights

Next we discuss the space maneuvers that can transfer a space vehicle from one circular orbit to another.

Suppose we need to launch a space vehicle from the orbital station into a circular orbit whose radius is different from that of the space station. After remaining in this new orbit for a while, the space vehicle is to return to the orbital station and dock to it. What maneuvers must be planned to execute this operation? What jet impulses are required for optimal maneuvers? What characteristic velocities must the rocket engine provide?

Designing such transitions between different circular orbits can be related to interplanetary space journeys. The orbits of the planets are almost circular and to a first approximation they lie in the same plane. In a sense, planets are stations orbiting the Sun. Sending a space vehicle from one planetary orbit to another differs from the problem suggested above only in that the planets (unlike actual stations) exert a significant gravitational pull on the space vehicle. But since the masses of the planets are small compared to the mass of the Sun, the gravitational field of a planet is effective only in a relatively small sphere centered at the planet. (Section 11.12 discusses details.) Outside this *sphere of gravitational action of the planet*, the motion of a space vehicle (relative to the heliocentric reference frame) is essentially a Keplerian motion governed by the Sun. In this sense the problem of interplanetary flights is quite similar to the problem to be discussed here. The only difference is that in the case of interplanetary flights the additional velocity needed to simulate a maneuver on the computer should be treated as the velocity with which the space vehicle leaves the sphere of gravitational action rather than the surface of the planet.

The most economical way to jump between two circular orbits (with respect to the required rocket fuel expenditure) is a semielliptic trajectory that grazes the inner orbit from the outside and the outer orbit from the inside. Such a transition is called *semielliptic* or *Hohman's transition* after W Hohman, a German scientist who was the first to suggest it for interplanetary flight.

As a particular example, we next consider the voyage of a spacecraft from an orbital station that moves around a planet in an inner circular orbit of radius $r_0$ to an outer circular orbit of radius 2 $r_0$. After remaining in this new orbit for a while, the spacecraft returns to the orbital station. Figure 5.10 illustrates the maneuvers. At point $P_1$ the space vehicle is undocked from the station and the on-board rocket engine imparts to the vehicle an additional velocity $\Delta v_1$ in the direction of the orbital motion. In order to acquire an apogee of $2r_0$ for the transitional semielliptic

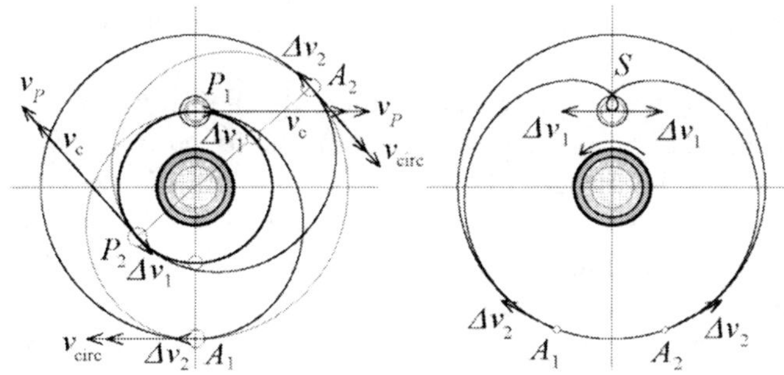

**Figure 5.10.** The semielliptic transition of a spacecraft to a higher circular orbit with subsequent return to the orbital station.

trajectory, the additional velocity $\Delta v_1$ must equal 0.1547 $v_c$, where $v_c$ is the orbital velocity of the station. The calculation of the required additional velocity $\Delta v_1$ on the basis of the laws of conservation of energy and angular momentum is given in section 10.9 'Space Rendezvous'. When the space vehicle reaches the apogee (point $A_1$) of the ellipse, a second tangential impulse $\Delta v_2$ is required to increase the velocity to the value $v_{circ}$, in order to place the space vehicle in the outer circular orbit.

An additional velocity of the same magnitude $\Delta v_2$, but directed opposite to the orbital velocity, is required to transfer the space vehicle to a semielliptic trajectory that can bring it back to the station. However, when the orbital station is to be the target, another important consideration is timing: the station must be at the right spot in its orbit at just the moment when the space vehicle arrives. Therefore the instant and the point $A_2$ at which the maneuver is carried out must be chosen properly in order that the space vehicle reach the perigee $P_2$ simultaneously with the station. To calculate a suitable time, we can use Kepler's third law. (Chapter 10 gives details.) To equalize the velocity of the space vehicle with the velocity of the station, one more rocket impulse is required. It is obvious that now the required additional velocity has the same magnitude $\Delta v_1$ as it does for the very first maneuver.

The right panel of figure 5.10 illustrates the motion of the space vehicle in the frame of reference associated with the orbital station. We note that between points $A_1$ and $A_2$ the space vehicle covers more than one revolution around the station in its retrograde relative motion, while in the planetocentric motion between the corresponding points $A_1$ and $A_2$ (left panel of figure 5.10) it covers less than one revolution.

## Questions and problems

1. **Semielliptical transitions between circular orbits.** Explain why the semielliptic trajectory (Hohman's transition) provides the most economical way (with respect to the required rocket fuel expenditure) to transfer a spacecraft from one circular orbit to another circular orbit of a different radius.
2. (*) **Characteristic velocities.** For the maneuvers discussed above in this section, calculate the values $\Delta v_1$ and $\Delta v_2$ of the additional velocity required to transfer the spacecraft between the initial circular orbit (orbit of the space station) and another circular orbit of a given radius along semielliptic transitional trajectories.

Express the values of the characteristic velocities in units of the circular velocity of the space station.

3. (***) **Chronometry of the maneuvers.** Calculate the moment $t_2$ (assuming $t_1 = 0$ for the first maneuver of departure from the orbital station) at which the spacecraft reaches the apogee $A_1$ of the first semielliptic transitional trajectory. At this moment $t_2$ the second maneuver should be executed. Express $t_2$ in units of the period $T_0$ of revolution along the original circular orbit (orbit of the station). Calculate also a suitable moment $t_3$ for the maneuver of transition to the semi-elliptic return trajectory, in order for the spacecraft to reach the point of tangency with the circular orbit of the station just at the moment when the orbital station arrives at this point. Calculate $t_3$ for cases in which the spacecraft stays in the outer circular orbit for (a) more than a half-revolution and (b) more than a whole revolution. What should the time $t_4$ be in cases (a) and (b) for the final maneuver of equalizing the velocities and docking to the station? Verify your calculations using the simulation experiment.
4. (**) **Transition to the opposite point of a circular orbit.** A spacecraft docked to the orbital station must be transferred after undocking to the opposite point of the same circular orbit, so that finally the station and the spacecraft move along the same circular orbit being separated from each other by the angular distance of 180° (that is, they should always be at the opposite ends of a straight line passing through the center of the Earth). What instantaneous maneuvers of the undocked spacecraft can provide such a transition? What maneuvers can bring the spacecraft back to the orbital station?

# Chapter 6

## Precession of an equatorial orbit

The shape of the Earth is only approximately a sphere. Our planet bulges slightly at the equator so that it is very nearly an oblate spheroid whose equatorial radius is about 21 km greater than its polar radius.

Hence the gravitational field of the Earth is not exactly a spherically symmetric field. For distances considerably greater than the Earth's radius, the gravitational field can be treated as if it were created by a massive sphere with an additional massive belt surrounding the Earth along the equator. By virtue of this belt, an additional term (called a *perturbation*) appears in the expression for the gravitational force experienced by a satellite, a term that decreases much more rapidly with increasing distance than does the unperturbed force.

The distortion of the Earth's gravitational field from spherical symmetry causes the actual orbit of a satellite to differ from an ellipse. The surprisingly simple Keplerian closed orbits vanish. The real trajectory is a complex curve, generally not closed and not lying in a plane. After a revolution, the satellite does not return to the same spatial point. However, because the distortions of the gravitational field are small, it is convenient to consider the satellite to be orbiting the Earth along a Keplerian ellipse whose parameters and orientation are continuously changing. Such an ellipse with gradually varying parameters is called an *osculating orbit*.

For a very distant satellite, the exact distribution of the Earth's mass is insignificant. In other words, to a first approximation the gravitational force behaves as if all the Earth's mass were concentrated at the Earth's center. This characteristic is reflected in the fact that the perturbation in the expression for the force falls of with the distance much faster than does the principal inverse-square term. For large enough distances from the Earth, this additional term is inversely proportional to the *fourth power* of the distance and thus goes to zero with increasing distance much more rapidly than does the inverse-square term.

Furthermore, the additional force of attraction by the massive equatorial belt generally is not directed toward the center of the belt. In other words, the actual

doi:10.1088/978-0-750-31100-7ch6  

gravitational field of the Earth is not exactly a central field. It possesses axial symmetry rather than spherical symmetry. Due to this lack of spherical symmetry the plane of a non-equatorial orbit gradually rotates in space. Since the angle between the plane of the orbit and the Earth's axis remains constant, the motion of the plane is a slow precession about the Earth's axis.

However, in the equatorial plane the perturbing force does point toward the center of the Earth. The motion of an equatorial satellite occurs under a gravitational force that is central but that has a dependence on distance which differs from that of a spherically symmetric planet. We can show with the help of the law of universal gravitation and the principle of superposition that for large enough distances $r$ from the center of the planet the force of attraction in the equatorial plane can be written as follows:

$$F_r = -G\frac{mM}{r^2}\left(1 + b\frac{R^2}{r^2}\right). \tag{6.1}$$

Here $R$ is the radius of the planet, $M$ is its mass and $G$ is the universal gravitational constant. A dimensionless parameter $b$ characterizes the degree of oblateness of the planet's shape. For example, if we assume that some mass $\Delta m$ of the total mass $M$ is concentrated in a thin equatorial belt surrounding the planet, parameter $b$ is positive and equals $(3/4)\Delta m/M$. (For an analytic derivation of expression (6.1) on the basis of the assumed model of the oblate planet see section 11.8) For the Earth, the deviation of the shape from a sphere is small, so that its oblateness is characterized by $b = 0.0016$ in the perturbational term of (6.1).

Since the force depends only on the distance $r$, the trajectory of an equatorial satellite is a plane curve, though generally not a closed one.

As mentioned above, at very large distances from the Earth the perturbation caused by the oblateness is negligible and the gravitational field decreases almost as the square of the distance (see (6.1)). A satellite at such distances follows a Keplerian orbit. However, if the eccentricity of this ellipse is large enough, that is, if the ellipse is highly flattened and its perigee is much closer to the Earth than its apogee, the influence of the Earth's distortion at perigee can be significant. The force of attraction of the planet near the perigee is greater than it would be in the absence of the perturbation.

Because of this additional attraction, at small distances from the planet the curvature of the trajectory is greater than in the unperturbed motion. The increased curvature in the perigee causes the major axis of the ellipse to turn in the direction of the orbital motion. We can imagine the evolution of the orbit as a slow continuous rotation of the major axis in the equatorial plane. This rotation is non-uniform. It slows down at large distances, when the satellite moves near the apogee and speeds up when the satellite passes through the perigee of the ellipse.

Furthermore, the osculating ellipses for different points of the actual trajectory differ not only by the orientation of their major axes, but also by other parameters. In particular, the major axis of the osculating ellipse is smaller at the apogee than at the perigee. However, in contrast to the orientation of the ellipse that steadily turns

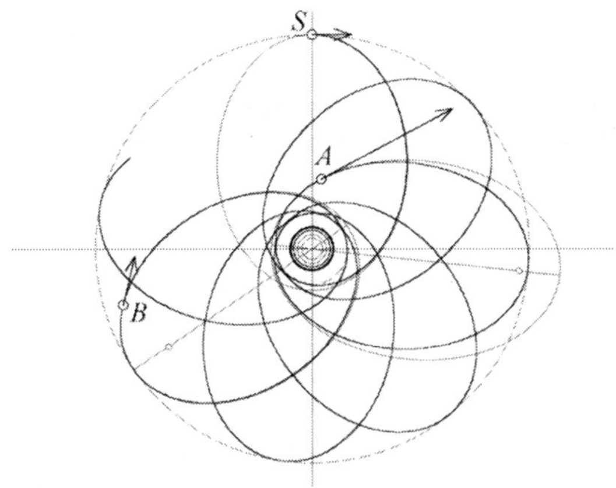


**Figure 6.1.** The precession of the orbit of an equatorial satellite orbiting an oblate planet.

in the same sense as the satellite is orbiting, variations in the other parameters occur periodically. As a result, for the complicated trajectory generated by this rotation and varying length of the major axis, the maximal and minimal distances of the satellite from the planet do not change from one loop to the next.

For a satellite orbiting the Earth, the major axis of its elliptical orbit turns only through a very small angle during one revolution of the satellite. However, the effect accumulates after a large number of revolutions around the Earth. To make the precession of an elliptical orbit easily observable in the computer simulation, we enter exaggerated values of the dimensionless parameter $b$ that characterizes the oblateness of the planet (up to $b \approx 1$ instead of $b = 0.0016$) in the expression for the gravitational force, (6.1).

The effect of precession is more evident for orbits with large eccentricities. To observe such flattened orbits in the simulation experiment, we choose an initial velocity which differs considerably from the circular velocity for the initial altitude. The trajectory of a satellite looks like a multi-petalled flower whose leaves gradually fill out the annular (ring-shaped) region of the equatorial plane enclosed between the two concentric circles corresponding to maximal and minimal distances of the satellite from the planet. These circles are shown by dashed lines in figure 6.1. Generally, this filling of the annular domain is uniformly dense. However, when the period of rotation of the major axis is an integer or rational multiple of the period of revolution (more exactly, multiple of the period of radial motion) of the satellite, the orbit is closed and consists of several loops.

The program allows us to draw the osculating ellipse on the screen for any point of the actual complex trajectory of the satellite. To obtain the curve during the simulation, we click the command button 'Pause' at the appropriate moment. If the option 'Osculating ellipse' in the menu item 'Options' has been selected, the program calculates parameters of the ellipse for the current values of the position and velocity of the satellite and shows the theoretically determined ellipse on the screen. In figure 6.1 such ellipses are shown (by thin lines) for three points ($S$, $A$ and $B$) of the trajectory.

By clicking the button 'Go' we resume the simulation and can compare the actual trajectory with the curve analytically predicted for the unperturbed motion.

If the point for drawing the osculating ellipse is chosen rather far from the planet, the actual motion follows this ellipse almost exactly until the satellite approaches the perigee (the ellipses for points *S* and *B* in figure 6.1). Near the perigee the actual motion deviates considerably from the ellipse and for the next loop of the trajectory the motion occurs approximately along an ellipse whose major axis is turned in the direction of the orbital motion through a considerable angle. If the osculating ellipse corresponds to some point that is not far from the planet (point *A* in figure 6.1), the actual motion occurs along the ellipse only in the near vicinity of this point and very soon deviates considerably from the predicted curve.

The deviation of the actual trajectory from the osculating ellipse appears very soon if the moment for drawing the ellipse is chosen when the satellite approaches the planet, although the satellite followed this ellipse well on its way toward the planet. In this case the ellipse gives a good approximation for the trajectory of actual subsequent motion during some time interval after the satellite passes through the corresponding point of the trajectory.

The osculating ellipse in the simulation experiment may be obtained by a different method. If, before starting the simulation, we select the option 'Unperturbed curve' in the menu item 'Options', the program numerically calculates the further unperturbed motion for the next period each time the 'Pause' button is clicked. We emphasize that the displayed unperturbed motion is not assumed by the computer to occur along an ellipse: the program calculates the unperturbed motion only on the basis of the fundamental equations of motion just as it does for the actual motion. At the moment the 'Pause' button is clicked, the subsequent motion of the satellite (for a period) is assumed to occur in the unperturbed spherically symmetric gravitational field, with the initial conditions equal to the current values of the position and velocity. If both options ('Osculating ellipse' and 'Unperturbed curve') are selected, after the 'Pause' button is clicked, the program first draws the analytically predicted osculating ellipse and then the unperturbed curve obtained by the numerical integration of the equations of motion.

Among the possible equatorial trajectories in the perturbed gravitational field, for any initial distance there always exists a closed circular orbit. Since the centripetal force is greater than for the case of an undistorted planet, the velocity of a satellite in this circular motion must exceed the circular velocity which is characteristic of the undistorted planet of the same mass. If we imagine that the distortion suddenly vanishes, such a circular satellite would move after this moment along an elliptical orbit with the perigee at the initial point, where this ellipse grazes the circle. In other words, for the circular motion of a satellite under perturbed gravitational force, the varying osculating orbit is an ellipse whose major axis makes exactly one revolution during one period of the perturbed circular motion. The satellite is always at the perigee of the osculating ellipse.

It is also interesting to consider the hypothetical case of a planet with the opposite distortion of mass distribution, namely a planet whose shape is a prolate spheroid. In this case the additional term in the expression for the gravitational force is negative

($b < 0$ in (6.1)), since the force of attraction at small distances is less than for an undistorted planet of the same mass. To explain this, we can imagine a planet with polar 'hats' of snow and ice (as if some of the planet's mass were transferred to its poles), so that the gravitational field is created by a massive sphere and by additional point masses at its poles. For any outer point in the equatorial plane, the distance to the poles is greater than the distance to the center of the planet. The greater part of the planet's mass is transferred to the poles, and the greater the separation between these point masses, the smaller the resulting force of attraction acting on a satellite in the equatorial plane at a given distance from the center of the planet. (For an analytic derivation of expression (6.1) on the basis of the assumed model of the prolate planet see section 11.8.)

When an equatorial satellite passes through the perigee of its orbit, the gravitational attraction by the dilated planet is insufficient to provide the curvature of the unperturbed trajectory. Such straightening of the ellipse in its perigee is equivalent to a turn of the major axis through some angle opposite to the direction of the orbital motion.

Using the simulation program, we can draw the osculating unperturbed ellipse for any point of the actual trajectory. The ellipse is generated each time we click the button 'Pause' in the course of the simulation, provided the corresponding menu item in 'Options' is checked. The osculating ellipse is obtained, depending on the option chosen, either by the numerical simulation of motion under the unperturbed gravitational force, or by a theoretical calculation of the parameters of the elliptical orbit for the current values of the velocity and position in the actual motion. We can also choose both these options to verify the theoretical predictions by a numerical simulation.

If we generate the osculating ellipse at a time the satellite is moving away from perigee and is sufficiently far from it, the satellite in the subsequent perturbed motion follows this ellipse almost exactly until it again approaches the perigee, where its motion is subjected to the strongest of the perturbations. When the osculating ellipse is generated for a point that is close to the perigee, the osculating ellipse is not a good approximation of the subsequent perturbed motion.

We should keep in mind that the expression for the gravitational force generated by a distorted planet (used for the simulation of motion in the computer program) is approximate and valid only for rather large distances from the planet. For cases in which the trajectory closely approaches the surface of a significantly distorted planet, the adopted mathematical model is inaccurate.

## Questions and problems

1. **Gravitational field of an oblate planet.** For a planet whose shape is an oblate spheroid, why is the gravitational force exerted on a satellite in the equatorial plane at relatively small distances stronger than for a spherically symmetric planet of the same mass? What can you say about the gravitational force over the poles of the oblate planet? Is it greater or smaller than for the spherically symmetric planet of the same mass?
2. **Gravitational field of a dilated planet.** For a planet whose shape is a prolate spheroid, why is the gravitational force acting upon a satellite in the equatorial

plane smaller than for a planet with a spherically symmetric distribution of the same mass? Is this true for the gravitational force over the poles of the planet?

3. (**) **Inverse fourth-power fall-off of the additional force.** Prove that the additional term in the expression for the gravitational force created by an oblate or prolate planet at large distances is inversely proportional to the fourth power of the distance. Use the model of an equatorial massive belt for an oblate planet, and of point masses at the poles for prolate planet.
4. **Direction of the precession.** For an oblate planet, explain qualitatively why the precession of an equatorial orbit occurs in the same sense as the orbital motion, while for the prolate planet, the precession occurs in the opposite sense to the orbital motion of the satellite.
5. (*) **Maximal and minimal distances.** Prove on the basis of the laws of angular momentum and energy conservation that the maximal and minimal distances of the equatorial satellite from an axially symmetric distorted planet do not change from one revolution to another in spite of the variations of the major axis of the osculating orbit.
6. **Circular orbit around the non-spherically symmetric planet.** Calculate the value of velocity of the equatorial satellite orbiting an axially symmetric planet of mass $M$ along a circle of radius $r$ if the distortion of the planet is characterized by a definite value of the dimensionless parameter $b$.
7. (*) **Osculating ellipse for a circular orbit.** Explain why for a circular motion of an equatorial satellite orbiting an axially distorted planet the osculating orbit is a rotating ellipse. What is the angular velocity of this rotation? In which case is the satellite always found at the perigee, and in which case at the apogee of the osculating ellipse?
8. (*) **Maximal and minimal distances from the center.** A satellite is launched with some initial velocity $v$ in the equatorial plane of a distorted planet transverse to the radius vector whose length is $r$. Calculate the maximal and minimal distances of the satellite from the center of the planet. The distortion of the planet is characterized by a dimensionless parameter $b$.
9. **A closed multi-petalled orbit.** Using trial and error, find the initial conditions that produce a precessing orbit that becomes a closed curve after several revolutions.

# Chapter 7

## Binary stars—the two-body problem

The motion of two bodies coupled by gravitational forces is simulated in the program 'Double Star'. When the bodies have comparable masses, such a physical system is a model of a binary star whose components revolve about a common center of mass. The program shows that in the inertial center-of-mass reference frame, the bodies trace synchronously homothetic Keplerian orbits.

In the preceding programs of the package `PLANETS AND SATELLITES` the simulation of motion under the action of gravitational forces is carried out under the assumption that the mass of the central body is much larger than the mass of the orbiting body. Hence the more massive central body (the Sun in the problem of planetary motion or the Earth in the problem of satellites orbiting the Earth) can be approximately treated as stationary and the problem is reduced to the investigation of orbital motion of the less massive body in the gravitational field of the other.

In the general case, when the masses of the interacting bodies are comparable, such an approximation is not possible because neither of the bodies is in fact stationary relative to their inertial center-of-mass frame. Gravitation is mutual: if the Earth pulls on the Moon, the Moon pulls on the Earth. That is, the more massive central body is also forced to move under the gravitational pull produced by the other orbiting body and this motion of the central body influences in turn the motion of the orbiting body, and so on. To be precise, we should avoid saying that the Moon orbits the Earth. In fact, the Moon and the Earth orbit each other, revolving around their common center of mass. Therefore, it is necessary to take into consideration the motions of both interacting bodies, the treatment of which is called the *two-body problem*.

The forces of interaction between the bodies, in accordance with Newton's third law, are equal and opposite. This fundamental law is valid for every known interaction between bodies, independent of the physical nature of the interaction. In particular, it holds for gravitational interaction. The universal character of Newton's third law is associated with the conservation of momentum in an isolated physical system and, more generally, with the homogeneity of physical space.

doi:10.1088/978-0-750-31100-7ch7

Applying Newton's third law to the system of two interacting bodies, it is possible to reduce the two-body problem to the problem of motion of a single body with a mass $\mu$, the *reduced mass*, whose value is given by $\mu = m_1 m_2/(m_1 + m_2)$. This body moves under the action of a central force equal to the force of interaction between the actual bodies (irrespective of the physical nature of the interaction). It can be shown that the solution to this problem describes their *relative* motion, that is, the motion of one of the bodies in the (non-inertial) reference frame associated with the other body. (See section 11.9 for details.) When the interaction is universal gravitation (whose force decreases with the square of the distance between the bodies), an exact analytical solution is available which shows that this relative motion obeys Kepler's laws.

After the problem of the motion of one body relative to the other is solved, we can easily find the motion of each of the bodies relative to their inertial center-of-mass frame. In this frame, the two bodies move in conic sections about one another, with the focus for each trajectory at the stationary common center of mass of the system. Knowing the motion of the bodies relative to the center of mass, we can then find their motion relative to an arbitrary inertial reference frame, taking into account that the motion of the center of mass of the whole system with respect to any inertial reference frame is uniform rectilinear motion.

The simulation program 'Double Star' of the package `PLANETS AND SATELLITES` illustrates the two-body problem for the special case of a binary star whose components may have comparable masses and move under mutual gravitational attraction. To understand the peculiarities of the problem better, we should first explore the motion in the inertial reference frame associated with the center of mass of the system (the corresponding item in the menu 'Options' should be selected).

Before the simulation we can enter the ratio of the masses of the stars and their velocities. (We need not enter the initial distance between the stars because in the program the distances are measured in units of this initial distance.) It is assumed that at the initial moment the velocities of the stars (in the center-of-mass reference frame) are perpendicular to the straight line joining the stars. Since in this reference frame the magnitudes of the velocities are inversely proportional to the masses of the stars, it is sufficient to enter the velocity for only one of the stars. We must express it in units of circular velocity. That is, if we enter 1 unit for the transverse velocity, the components of the binary star will move synchronously along concentric circular orbits. In this case the distance between the stars remains constant during the motion. If we enter 0.5 for the initial velocity, each component initially moves at half the circular-orbit speed. When the initial transverse velocity is less than the circular velocity, the stars initially are at the greatest distance from one another, and vice versa.

The actual value of the velocity (say, in km $s^{-1}$ instead of the circular-orbit speed as the unit of velocity) depends on the masses of the double star components and the distance between them.

The relative motion of the bodies in the two-body problem is equivalent, as we already mentioned, to the motion of a single body under a central force that equals the force of interaction between the bodies. In the special case of a double star, this

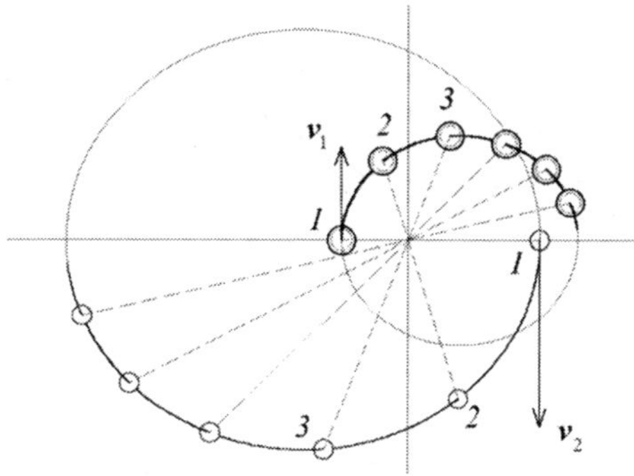


**Figure 7.1.** The trajectories traced by double star components in the center-of-mass reference frame. The simultaneous positions of the bodies are marked by the same numbers.

force is the inverse-square Newtonian gravitational force. Therefore the relative motion of the double star components is a Keplerian motion studied in the preceding programs.

The center of mass of the system is located on the straight line joining the stars and divides this line in a constant ratio inverse to the ratio of the masses. In the inertial reference frame associated with the center of mass (figure 7.1), both stars move synchronously around geometrically similar (homothetic) elliptical orbits whose common focus is located at the center of mass of the system. That is, both orbits have the same eccentricity. (In particular, the orbits may be concentric circles.) The linear dimensions of these similar orbits are inversely proportional to the masses of the stars. One of the foci of each ellipse is located at the stationary center of mass of the system and the major axis of each ellipse passes along the same straight line.

As the stars move, they are always at the ends of a rotating straight line that passes through the stationary common focus of their orbits located at the center of mass. (This rotating line is shown for different instants by dashed lines in figure 7.1). Therefore the trajectory of the relative motion (the motion of the smaller star with respect to the larger one or vice versa) is an ellipse homothetic to the ellipses traced by the stars in the center-of-mass frame. Any linear dimension (e.g. the major or minor axis) of this ellipse of the relative motion equals the sum of the corresponding linear dimensions of the ellipses traced by both stars relative to the center of mass.

Since the force of gravity between the stars lies at each instant along the line joining the stars, the force vectors are directed through the center of mass. Hence the angular momentum of each star relative to the center of mass does not change during the motion, and Kepler's second law is satisfied. That is, the dashed lines (shown in figure 7.1) sweep out equal areas of each ellipse in equal times. Therefore for elliptic motions the line joining the stars rotates non-uniformly. The rotation is fastest when the distance between the stars is minimal.

In order to explain why the motions of the stars relative to the center of mass obey Kepler's first law (i.e. occur along ellipses), we can show that each of the stars coupled by mutual gravitation can be treated as moving not under the pull of the

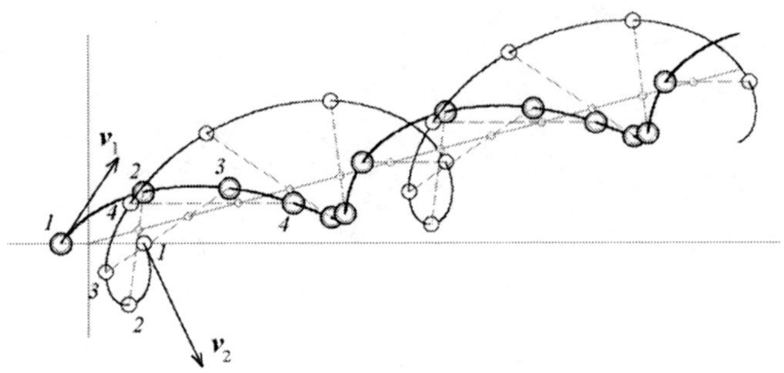


**Figure 7.2.** Trajectories of a double star components in an arbitrary inertial reference frame. The simultaneous positions of the bodies are marked by the same numbers.

other moving star, but rather in a stationary central gravitational field whose strength diminishes as the square of the distance from the center of mass. The accelerations of the stars in these synchronous motions relate as their distances from the center of mass.

The program ‘Double Star’ allows us to open an additional window in order to simultaneously display the relative motion of the double star components. This (non-inertial) frame of reference can be associated with any of the stars, depending on the item chosen from the menu ‘Frames’. We can see in the simulation that one of the components moves around the other in an ellipse homothetic to the ellipses traced by the components of the double star in the inertial center-of-mass frame.

To observe the motion of the double star components in an arbitrary inertial frame of reference, we open the panel ‘Input’, choose the option ‘Arbitrary reference frame’, and enter the velocity of the center of mass with respect to the reference frame in which the motion is to be displayed. The simulation program deals only with two-dimensional planar systems. (This limitation is caused solely by the difficulties of reproduction and visual perception of three-dimensional motion on a two-dimensional computer screen.) Therefore the velocity vector of the center of mass must lie in the plane of the orbital motion of the stars. The magnitude of this velocity must be specified in units of the transverse circular velocity of the less massive component, measured with respect to the center-of-mass frame. The direction of the velocity of the center of mass must be indicated by entering the angle (in degrees) that this velocity makes with the perpendicular to the line joining the double star components at the initial moment.

Figure 7.2 shows trajectories of the double star components in an arbitrary inertial frame of reference. The simultaneous positions of the stars are joined by dashed lines. It is clear that the center of mass (shown by a small circle in the figure) moves uniformly along a straight line. The stars themselves move non-uniformly along complicated wavy or looped trajectories that are generated by the superposition of their periodic Keplerian elliptical motions around the center of mass and the uniform rectilinear motion alongside the center of mass.

A double star is called a *visual binary* if its components are far enough apart to be seen separately through an optical telescope. By measuring the period of revolution and the orbit of relative motion, we can determine the sum of the masses of the stars. To determine the mass of each of the stars separately, we must measure their orbits traced around the center of mass. (The dimensions of these orbits are inversely proportional to the masses of the stars.) Currently, thousands of visual binaries with orbital periods from several years to many thousands of years have been recorded.

In an *astrometric binary*, one component is too faint to be seen and its presence is inferred from the perturbations in the visible motion of the other component on the background of remote stars. In this way the first *white dwarfs* (compact stellar objects formed as the end products of the evolution of stars of relatively low mass) were discovered. The complicated wavy trajectory of Sirius measured relative to the stars gave evidence of the presence of an invisible satellite that was afterward discovered visually.

In a *spectroscopic binary*, the stars are so close that they usually cannot be resolved by a telescope, but their relative motion can be detected by periodic variations in the observed spectrum. These variations are caused by different Doppler shifts of lines common to the spectra of both stars. At conjunctions both stars move perpendicularly to the line of vision and their spectral lines coincide. After a quarter of the period, one of the components moves towards the observer at its greatest speed while the other recedes from the observer, and the lines in their spectra are shifted oppositely in proportion to their speeds along the line of vision.

## Questions and problems

1. (*) **Relative and 'absolute' motion.** Assuming that one of the stars of a binary system moves around the other in a Keplerian ellipse, prove that relative to the center of mass the stars trace homothetic elliptical orbits with the common focus at the center of mass. What is the ratio of the linear dimensions of the ellipses?
2. (*) **A simple double star and scaling principles.** The components of a double star have equal masses and move in a circle. By what factor does their period of revolution change if the spatial scale of the system (the distance between the components) is increased by a factor of 4 (holding the motion circular and the masses constant)?

   By what factor does the period change if the masses are increased by a factor of 2 (holding the spatial distance between the components constant)?
3. (**) **Scaling for an arbitrary double star.** Answer the previous questions for a double star with non-equal components that move around the center of mass along elliptical orbits.
4. (*) **Effective stationary gravitational field.** Show that in the two-body problem each of the bodies coupled by mutual gravitation can be treated as moving not under the pull of the other moving body, but rather in a stationary central gravitational field whose strength is inversely proportional to the square of the distance of the body from the center of mass of the system. Prove that the accelerations of the bodies in this motion relate as their distances from the center of mass: $a_1/a_2 = r_1/r_2$.

# Chapter 8

## Three-body systems

The most fascinating phenomena of celestial mechanics are revealed in investigating the motion of a system of three or more bodies attracted to one another by gravitational forces. The systems of three and more interacting bodies are simulated in several programs of the package `PLANETS AND SATELLITES`. Among these are: the motion of a satellite orbiting a planet that is orbiting a star; the motion of a planet in a double-star system; and the motion of several planets orbiting a single star. The programs allow us to observe and study many fascinating trajectories of three-body and many-body motion that challenge our intuition and delight the eye.

### 8.1 The restricted three-body problem

The programs 'Planet with a Satellite' and 'Double Star with a Planet' deal with a *restricted three-body problem*, in which the mass of one of the bodies is negligible compared to the masses of the other two. In this case we can ignore the influence of the least massive body (the 'light' body) on the motion of the other two. Hence the motion of the two massive bodies (the 'heavy' bodies) is exactly Keplerian, as described above in the section 'Binary stars' and illustrated by the corresponding simulation program 'Double Star'. It is shown there that, with respect to the inertial frame of reference associated with the center of mass of the system, the bodies trace homothetic Keplerian orbits, whose linear dimensions are inversely proportional to the masses of the heavy bodies. The common focus of the orbits is located at the stationary center of mass of the system and the radius vectors from the center of mass to each star trace out equal areas in equal times.

Our current interest in the restricted three-body problem is the motion of the light body of negligible mass. This motion occurs under the forces of gravitation created by the two heavy bodies whose motion is already known. Even for the restricted three-body problem there is no general analytic solution. That is, there is no solution that determines the motion of the light body under arbitrary initial conditions. The absence of an analytic solution to the differential equations of motion for so

doi:10.1088/978-0-750-31100-7ch8

simple a system is probably related to the extreme complexity of possible motions of the system. For some values of parameters of the system and/or initial conditions, the motion of the light body is irregular, seemingly random (chaotic), in spite of the deterministic character of the problem. The chaotic behavior of nonlinear systems governed by simple deterministic laws is related to the extreme sensitivity of the differential equations describing the system to the initial conditions: a very small initial difference may result in an enormous change in the future state and long-term behavior of the system. Celestial dynamics gives one of the numerous examples of chaos in physics.

## 8.2 Managing the program 'Planet with a Satellite'

The simulation program 'Planet with a Satellite' allows us to obtain a numerical solution to the restricted three-body problem for arbitrary initial conditions and arbitrary values of parameters of the system, and to observe the motion directly on the computer screen.

The motion of the simulated system can be displayed either in the frame of reference associated with any of the two massive bodies, or in the center-of-mass reference frame. It is also possible to observe the motion simultaneously in two different reference frames.

We emphasize that this program can deal not only with a satellite orbiting a planet (the primary) that, in turn, orbits a star, but also with various physical systems whose behavior is described by the restricted three-body problem. For example, we can simulate the motion of an interior light planet in a double-star system, or the motion of a spacecraft while en route toward the Moon under the action of the gravitational forces of both the Earth and the Moon. Or we can investigate the lunar or solar gravitational perturbations in the motion of an artificial satellite orbiting the Earth.

Finally, the program can simulate the motion in those interesting exotic special cases for which the restricted three-body problem has exact solutions. With the help of the program we can investigate, for example, the motion of a satellite in the vicinity of the libration points in a system of two massive bodies that orbit one another under the forces of mutual gravitation. (The Earth and the Moon, or the Sun and Jupiter, give approximate realizations of such a system.)

To reproduce the simulations described below, we can avoid laboriously entering the parameters of the system and the initial conditions since we have the option of choosing a simulation from the menu item 'Examples', which contains a list of prepared situations. During the simulation we may change the reference frame in which the motion is displayed, or we can open two reference frames to simultaneously display the motion as it is seen by different observers.

However, we may also conduct a simulation experiment of our own design. To do so, we need to enter the parameters of the system and the initial conditions in the panel 'Settings'. For the two massive bodies, we need to state the ratio of their masses and the initial velocity that determines their relative motion. This velocity is directed transversely (that is, perpendicularly to the line joining the bodies).

Therefore we should enter only its magnitude expressed in units of the circular velocity (the velocity for which the relative motion of the bodies and their motions in the center-of-mass reference frame are circular). That is, if 1 unit is entered, the bodies move in circles, and if 0.5 is entered, each body initially moves at half the speed each must have to sustain a circular orbit.

Next we enter the initial position and velocity of the satellite. To do so, we can choose one of the following frames of reference: that associated with the planet (which we later call it the geocentric reference frame), that associated with the star (heliocentric), or the inertial reference frame associated with the center of mass of the system. Which of the reference frames is more convenient depends on the problem we are investigating. The choice is made by clicking the appropriate option button.

The initial position of the satellite is defined in terms of its distance from the planet, its primary (or from the star, or from the center of mass, depending on the frame of reference that has been chosen), measured in units of the initial distance between the star and the planet, and the angle which the radius vector of the satellite drawn from the planet (or from the star, or from the center of mass) makes with the straight line directed from the star toward the planet.

The initial velocity of the satellite is indicated in the same way. We first enter its magnitude. If we use the geocentric reference frame, the magnitude of the initial velocity can be expressed either in units of the unperturbed circular velocity of the satellite for its initial distance from the planet (the velocity for which the satellite moves in a circular orbit about the planet in the absence of the gravitational influence of the star), or in units of the circular velocity of the planet (for the initial distance between the planet and the star). The choice in units is made by clicking the corresponding option button which appears when the box for the input of the velocity (or the scroll-bar) gets focus. If the heliocentric or the center-of-mass reference frame is used, the magnitude of the satellite's initial velocity must be expressed in units of the circular velocity of the planet in the chosen frame of reference.

Finally, we can enter the values for the radii of the star and the planet (measured in units of the initial distance between the star and the planet). These values may be important for situations in which the trajectory of the satellite passes very close to the star or to the planet. Depending on the radii of these celestial bodies, the satellite may either safely bypass the body or crash onto its surface.

## 8.3 Satellites of a planet that orbits a star

In the Solar System the masses of the planets are very small compared to the mass of the Sun. Therefore, the gravitational attraction of the planets to the Sun is much more important than the mutual attraction of the planets. The latter causes only small deviations from Kepler's laws.

If we consider a star with a single planet, its orbital motion is exactly Keplerian. Now let us imagine that a satellite of a negligible mass is orbiting this single planet which, in turn, is orbiting the star (figure 8.1). Earlier in our study of a satellite's

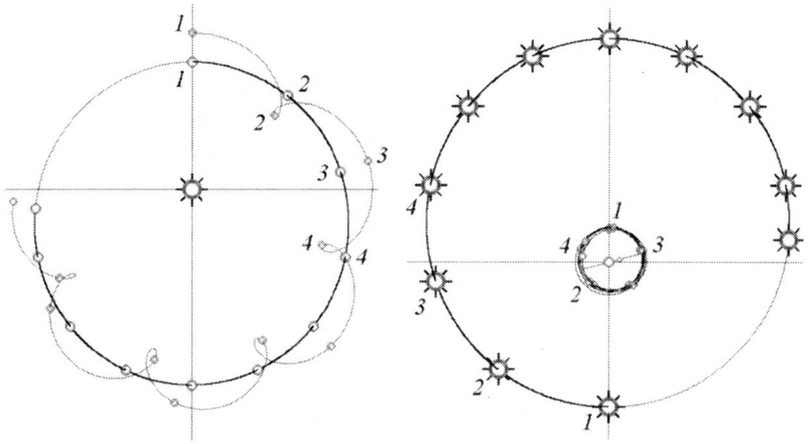


**Figure 8.1.** The trajectory of a satellite orbiting a planet in the heliocentric (left) and geocentric (right) frames of reference. Coinciding numbers refer to the same instants of time.

orbital motion around a planet we did not take into account the influence of the star. However, the satellite is subjected to the gravitational pull of the star as well as to the pull of the planet. For example, the force of attraction of the Moon to the Sun is greater than the force of attraction of the Moon to the Earth. In such situations, why do we say that the satellite orbits the planet, when the force of its attraction to the star is greater than the force of its attraction to the planet?

The point is that when we consider the motion of a satellite, we refer it to the *non-inertial frame of reference* associated with the planet. The planet itself is subjected to the gravitational pull of the star. In the gravitational field of the star the planet and its satellite acquire almost equal accelerations. Hence the influence of the star on the motion of the satellite relative to the planet is rather moderate and we can consider this relative motion to be Keplerian to a first approximation. The important factor is not the gravitational field of the star by itself, but rather the *inhomogeneity* of this field over spatial distances of the order of magnitude of the separation between the planet and the satellite. A more detailed discussion of the subject can be found in section 11.13.

The simulation program 'Planet with a Satellite' allows us to investigate the general character and peculiarities of this motion. Figure 8.1 depicts the screen image obtained in the simulation of the satellite that moves in an almost circular, low orbit. The simultaneous positions of the bodies on their trajectories are marked by equal numbers.

The complicated looping trajectory traced by a satellite in the heliocentric frame of reference (the frame centered on the star, shown in the left panel of figure 8.1) is explained by the addition of two rather simple motions: the revolution of the planet in a large ellipse around the star plus the simultaneous revolution of the satellite in a small, almost circular orbit around the planet. The right panel of figure 8.1 clearly

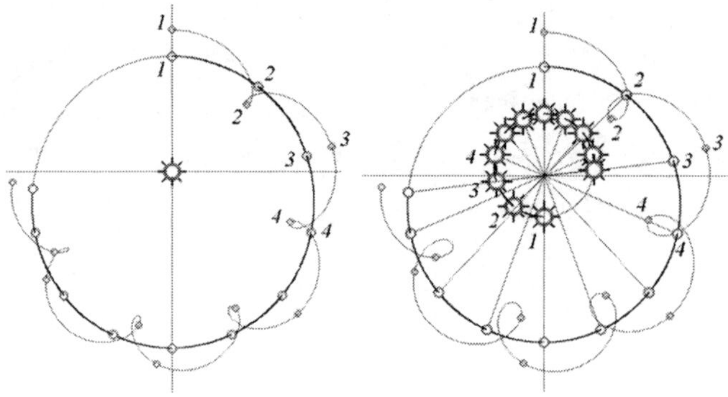

**Figure 8.2.** The trajectory of a satellite orbiting a planet in the heliocentric (left) and in the inertial center-of-mass (right) frames of reference. Equal numbers refer to corresponding instants of time.

shows that the geocentric motion of the satellite indeed occurs along a simple almost Keplerian orbit which is only slightly perturbed by the star. We see that the influence of the massive star on the orbital motion of the satellite around the planet is actually much less effective than the influence of the planet on the heliocentric motion of the satellite. Instead of a simple elliptical orbit (similar to the heliocentric orbit of the planet) we observe a complicated curve of loops or waves.

To make the peculiarities of the motion more obvious, we have chosen for the simulation values of parameters that are somewhat unrealistic for a real planetary system: the mass of the planet is almost one third of the mass of the star and so this simulation applies more reasonably to the motion of a light inner planet in a double-star system. However, the difference between the two situations (a satellite orbiting a planet orbiting a star versus a planet orbiting one of the components of a binary star) is quantitative rather than qualitative. If the mass of the second body is comparable to the mass of the heaviest one (the ratio of their masses is about 1/3 in the simulation shown in figure 8.2), the latter cannot be considered as stationary. Hence its reference frame (the 'heliocentric' frame in the simulation), as well as the 'geocentric' frame associated with the second massive body (the 'planet'), is non-inertial. Figure 8.2 shows the same motion of the three bodies both in the heliocentric reference frame and in the inertial center-of-mass frame. In the inertial reference frame both massive bodies move synchronously in homothetic ellipses with a common focus at the center of mass of the system.

In the simulated system of interacting bodies the forces of gravitational interaction depend only on the distances between the bodies. This character of the forces implies that the total mechanical energy of the system is conserved during the motion and that the equations of motion satisfy the symmetry with respect to the time reversal, that is, with respect to the replacement $t \to -t$. The symmetry of time reversal means that for any possible motion of the system there exists a symmetrical motion in which the bodies of the system pass through the same spatial points in reverse order with opposite velocities.

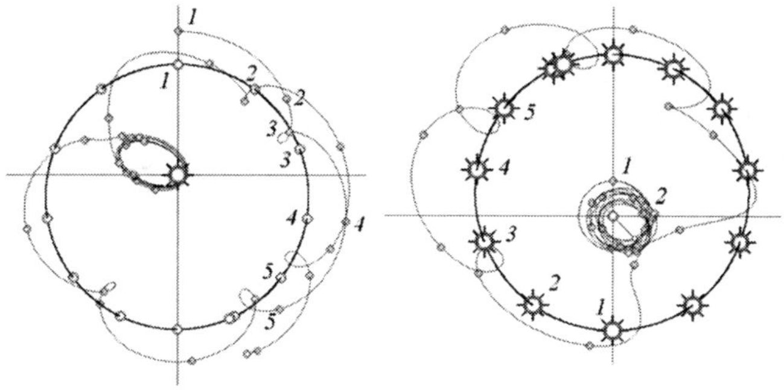

**Figure 8.3.** The trajectory of a satellite orbiting a planet and star in turn in the heliocentric (left) and in the geocentric (right) frames of reference. Equal numbers refer to corresponding instants of time.

The program 'Planet with a Satellite' allows us to verify the reversibility of motion experimentally. Under the menu 'Options' there is an item 'Reverse Velocities'[1]. When we choose it, the program instantly reverses the directions of the velocities of all the bodies. Then we can watch how the bodies move backward along the same trajectories and the system evolves toward the initial configuration. We note that the program does not store the previous positions of the bodies in its memory: after we click the item 'Reverse Velocities', the program simply continues to integrate numerically the same equations of motion and we see clearly that the bodies pass backward through the same positions. We can even observe the reversed motion for negative time values after the system passes through the initial configuration. During the simulation, we can reverse the velocities several times.

In the situations we have considered (the satellite of a star-orbiting planet or the planet in a double-star system), the motion of the light body can be *stable* in the sense that the satellite orbits its primary indefinitely. However, various *unstable* motions of the satellite (figure 8.3) may also occur. Depending upon the parameters of the system and on the initial conditions, these can terminate in the satellite falling into the primary or the companion, or the satellite eventually being ejected from the system. For some values of the parameters, when the orbit of the satellite approaches the boundaries of the sphere of gravitational action of the planet (section 11.12 provides details about this sphere), the geocentric motion of the satellite is strongly perturbed by the gravitational field of the star. It may happen that after several revolutions about the planet the gravitational attraction of the star pulls the satellite from the planet's 'embrace' and the satellite becomes an independent planet orbiting the star along an almost elliptical Keplerian orbit that is slightly perturbed by the planet. Such a case is illustrated in figure 8.3, showing the motion of the system both in the heliocentric and geocentric frames of reference.

It may happen that such a satellite lost by the planet, after several independent revolutions about the star, is again captured by the planet. In figure 8.3 such a

[1] Right-clicking the mouse anywhere in the window invokes a pop-up menu with several items. Among them there is also the item 'Reverse Velocities'.

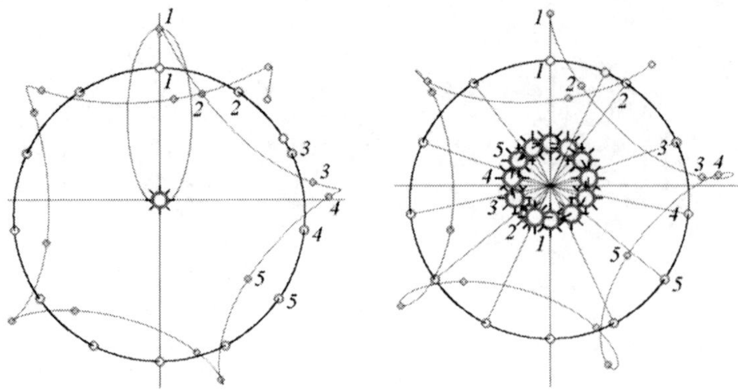

**Figure 8.4.** The trajectories of a satellite with a retrograde revolution around a planet in the heliocentric (left) and in the center-of-mass frames of reference.

'restitution' occurs after approximately a 'year' (one revolution of the planet around the star) of the satellite's independent existence. Similar exchanges of the satellite with the planet and the star in this 'game of space basketball' may be repeated many times. However, these extraordinary space voyages of the satellite eventually end with its falling into the planet or star, or by its ejection from the system. Similarly, in a double-star system, a planet that periodically makes transitions between orbits around each of the components cannot do so indefinitely. Sooner or later it crashes or is ejected. In spite of the simple deterministic laws that govern these three-body systems, their long-term behavior is seemingly random (chaotic) and unpredictable because of the extreme sensitivity to the initial conditions: a very small variation in the initial data may make an enormous change to the future state of the system.

The chaotic behavior of a non-linear system (governed by simple deterministic laws), which we observed in the above discussed examples, is related to the extreme sensitivity of the differential equations describing the system to the system parameters and initial conditions. Celestial dynamics gives one of the numerous examples of chaos in physics. We may suppose that in this case the absence of analytical solutions probably reflects the complexity of the possible motions of the system rather than the weakness of the analytic capability of the mathematics.

Figure 8.4 illustrates the interesting situation in which the satellite revolves about the planet in a direction opposite to that of the planet about the star. (Its revolution is said to be retrograde, as opposed to direct.) To simulate this situation, we choose the opposite direction for the satellite's initial velocity (in the geocentric frame of reference). The narrow stretched ellipse shows the Keplerian orbit along which the satellite (with the same initial position and the same heliocentric initial velocity) would move around the star in the absence of the planet.

Retrograde revolution is more stable against perturbations from the gravitational field of the star than is direct revolution of the satellite around the planet along an orbit of the same radius. This increased stability is explained by a greater velocity of the satellite relative to the star in the region between the planet and star, where the satellite is closest to the star.

The examples above clearly show how varied and complex the motions of a rather simple system of three interacting celestial bodies can be.

## 8.4 Exact particular solutions to the three-body problem

The program 'Planet with a Satellite' allows us to demonstrate the motions described by curious exact particular solutions to the restricted three-body problem. These solutions correspond to possible amazingly simple Keplerian motions of all three bodies along conic sections. It is a real wonder that such an unexpectedly simple finite subset of motions falls out of the continuous set of tremendously complex general three-body motions.

In all cases in which the motion of individual bodies (coupled by mutual gravitational forces in a many-body system) occurs along conic sections, each body can be treated mathematically as moving not under the pull of the other moving bodies, but rather in a stationary central gravitational field whose strength is inversely proportional to the square of the distance of the body from the center of mass of the system. We note that under certain conditions this effective gravitational field can be stationary in spite of the fact that it is created by moving bodies.

However, as a rule such a regular motion of the third body (of a negligible mass) is unstable with respect to (arbitrarily small) variations in the initial conditions. In simulation experiments we cannot enter the absolutely precise values required for the desired motion. Moreover, the numerical integration of the equations of motion is performed with limited precision. Therefore we can observe a regular motion of the system only during a finite time interval, after which the motion of the third body transforms into chaotic orbiting of one of the two massive bodies that continue to move in conic sections.

### 8.4.1 A system with equal masses of heavy bodies

We start our analysis of exact particular solutions to the many-body problem with the simplest possible example. When a pair of heavy bodies have equal masses, the restricted three-body problem has an evident exact solution, provided the third body (of negligible mass) is placed exactly halfway between the members of the pair (at the center of mass of the system), and provided its velocity relative to the center of mass is exactly zero. Then, since the gravitational forces on the central body due to each member of the pair are equal and opposite, the central body remains at the center of mass. This simplest particular solution is a special case of the collinear interior Lagrangian equilibrium point (see section 8.4.3).

When a pair of heavy bodies have equal masses, the three-body problem has an evident *exact solution*, provided the third body is placed exactly halfway between the members of the pair (at the center of mass of the system) and provided its velocity relative to the center of mass is exactly zero. Then, since the gravitational forces on the central body due to each member of the pair are equal and opposite, the central body remains at the center of mass. This is true for arbitrary symmetric motions of the massive bodies, including cases in which they trace synchronously congruent

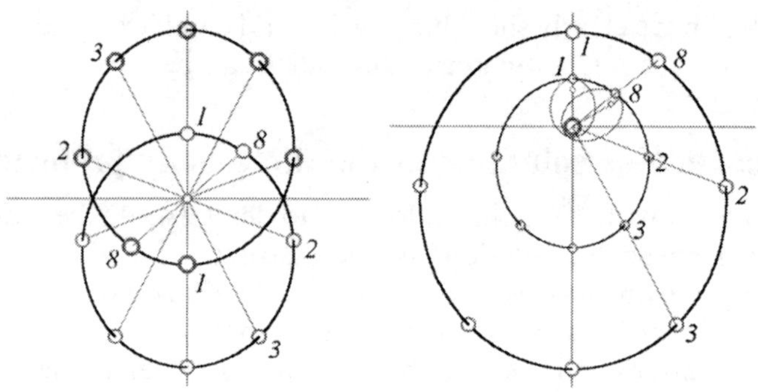


**Figure 8.5.** The three-body system (with two bodies of equal mass) whose motion is described by an exact particular solution.

elliptical orbits with the center of mass as their common focus (as depicted in the left panel of figure 8.5).

Moreover, this exact particular solution exists even if the third (central) body has a finite mass, that is, the solution holds for the *unrestricted* three-body problem. The only difference from the preceding case is that the third body, with an arbitrary mass, being placed halfway between the bodies of equal mass, influences their motion because of an additional gravitational pull: the net force on either one of the pair is the sum of two gravitational forces (pointing in the same direction), one from the central body and the other from the other member of the pair. However, the net force on either body of the pair is in this case also inversely proportional to the square of its distance from the central body (from the center of mass of the system).

Thus, for the symmetric initial configuration of the system, and equal and opposite initial velocities, the motion of each member of the pair about the center of mass of the system is Keplerian as if it moved in a stationary inverse-square central field. The symmetric configuration is preserved during the motion and the central body of arbitrary mass remains at rest in the equilibrium position (in the center-of-mass inertial frame of reference).

Although this exact solution is of no practical importance, its existence is interesting in principle and deserves discussion.

From the point of view of an observer on one of the paired bodies (that is, in the reference frame associated with this body), the central body moves in an ellipse homothetic to the ellipse traced by the other member of the pair (as depicted in the right panel of figure 8.5). The linear dimensions of this ellipse are half those of the relative elliptical orbit of the second paired body. The central body, moving synchronously with the second paired body, is always found at the exact center of the line joining the paired bodies. The simultaneous positions of the second paired body and the central body are marked with equal numbers. Two more small thin ellipses correspond to the unperturbed trajectories (for the initial position *1* and position *8*) along which the central body would move (in the frame of the first body) in the absence of the second body.

However, the equilibrium of the central body, as well as this simple motion of the whole system, is unstable. If the central body is even infinitesimally displaced from

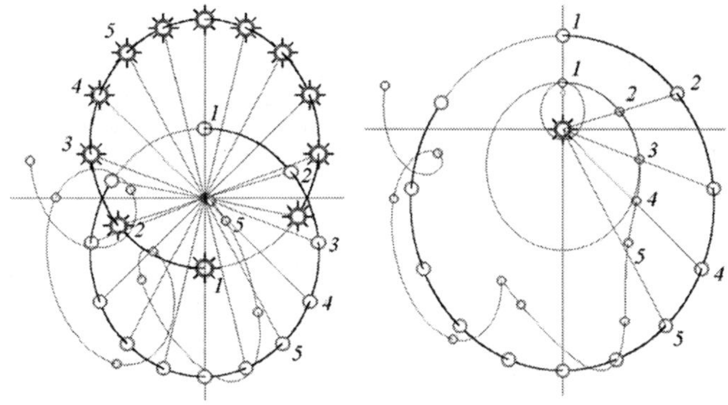


**Figure 8.6.** The instability of motion of a light body (placed halfway between two bodies of equal mass) that is described by an exact particular solution.

the center of mass and/or has an arbitrarily small velocity relative to the center of mass, it recedes from the center of mass with increasing velocity. Figure 8.6 illustrates this instability. In this example the initial distance of the central body from the first of the paired bodies is made slightly greater than its distance from the second. The central body, whose initial velocity relative to the center of mass is set at zero, remains for a while in the vicinity of the center of mass (as shown in the left panel of figure 8.6, which depicts the motion in the center-of-mass reference frame). However, from the initial moment, the central body begins to move toward the second (closer) paired body, about which it eventually orbits in an irregular trajectory that is highly perturbed by the first paired body. Relative to the first body, the central body deviates from the elliptical orbit that it follows in the exact analytic solution. This deviation becomes noticeable at the position marked *4* (after three 'months' of motion).

This solution also holds for arbitrary motions of the massive bodies, including cases in which they trace synchronously congruent open parabolic or hyperbolic trajectories with the center of mass as their common focus. Moreover, the existence of this exact particular solution is almost evident also in the case when the central body has a finite mass. That is, the solution holds for the unrestricted three-body problem, say, for the special case of an imaginary system of two planets of equal mass that orbit a single star.

### 8.4.2 Satellites at the triangular libration points

The restricted circular three-body problem has a set of interesting exact particular solutions. If two massive bodies orbit each other in circles, it can be shown that there exist five positions at which a much less massive body may be placed so that it orbits circularly about the center of mass of the system in the same plane and at the same angular speed as do the massive bodies. That is, the whole system rotates rigidly, as if the three bodies were the points of a solid rotating uniformly about the center of mass of the system. In other words, in the rotating reference frame associated with the line joining the massive bodies, the body of negligible mass is in equilibrium at

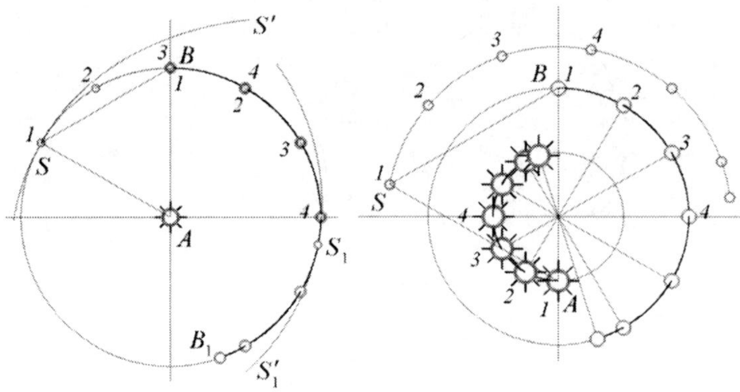


**Figure 8.7.** The motion of the satellite at a triangular libration point.

any of these positions. These five positions are called *libration points* (or Lagrange points). Lagrange points are formed by the combined gravitational forces of both massive bodies.

These exact solutions are of some practical interest in space dynamics because of the possibility (even if only in principle) of launching a stationary satellite located at one of the Lagrange points in the Earth–Moon system.

Three of the libration points are located on the line passing through the massive bodies (one point between the bodies). They are called *collinear libration points.* Each of the other two points is located at the apex of an equilateral triangle whose base is formed by the segment joining the massive bodies. These two points are called *triangular libration points.*

An analytic proof of the existence of the triangular libration points and the numerical calculation of the positions of collinear points are found in section 11.10. Here we describe the simulations in which these extraordinary exact solutions of the three-body problem are illustrated.

Figure 8.7 shows the stationary circular motion of a satellite at a triangular libration point of a system in which the ratio of masses $m_B/m_A$ equals 1/2. In the initial configuration the three bodies are located at the vertices of an equilateral triangle *BAS*. In the reference frame in which the body *A* is at rest (the left panel of figure 8.7), the body *B* moves along a circle with the center at *A*. The angular velocity of this motion depends on the distance *AB* between the bodies and on their masses. The initial velocity of the light body *S* equals the orbital velocity of the second body *B*.

If this second massive body *B* were absent, the satellite *S* with such an initial velocity would have moved under the gravitational pull of the body *A* along an elliptical orbit with the nearer focus at *A* because this initial velocity is greater than the circular velocity. The initial part *SS*′ of this osculating orbit (unperturbed by the second body *B*) is shown in the left panel of figure 8.7. However, the actual motion of the satellite *S* occurs along the same circular orbit as that of body *B*. The gravitational forces exerted on the satellite *S* by both massive bodies *A* and *B* together provide the necessary centripetal acceleration for the circular motion of the satellite $S$ with the same angular velocity as that of line $AB$.

Hence the initial equilateral configuration of the system is preserved during the motion. The sequential simultaneous positions of bodies $S$ and $B$ are marked in figure 8.7 by equal numbers. If body $B$ were suddenly to vanish at some instant (at point $B_1$ in figure 8.7), the further orbital motion of the satellite around body $A$ would occur along an ellipse with the perigee at point $S_1$. (This point $S_1$ is by 60° behind $B_1$.) A segment $S_1S_1'$ of this ellipse grazing the actual circular orbit is shown in figure 8.7.

The motion of the system in the inertial reference frame associated with the center of mass is illustrated in the right panel of figure 8.7. The equilateral triangle $ABS$ with the bodies at its vertices rotates uniformly as a whole (as a rigid body) about the center of mass.

Thus, in the reference frame associated with one of the massive bodies, say $A$, the other two bodies are revolving (clockwise in figure 8.7) along the same circular orbit, one of the bodies (the satellite $S$ in figure 8.7) lagging behind the other body ($B$) by an angle of 60°. (For the other triangular libration point, the satellite moves along the same circular orbit, in front of the other body by 60°.)

The triangular libration points are stable only if the ratio $m_B/m_A$ of the mass $m_B$ of one of the massive bodies to the mass $m_A$ of the other is small enough ($m_B/m_A < 0.04$). Since the mass of the Earth is approximately 81.3 times greater than that of the Moon ($m_B/m_A = 0.0123$), for the Earth–Moon system these points are stable.

In the Solar System, stable triangular libration points are formed by the combined gravitational forces of the most massive planet, Jupiter, and the Sun. There are two groups of asteroids (named Greeks and Trojans) that are trapped at Jupiter's leading and trailing triangular Lagrange points and move around the Sun synchronously with the planet.

For the system with $m_B/m_A = 1/2$, whose motion is displayed in figure 8.7, the triangular libration points are unstable. This instability in the motion of a satellite in the vicinity of the triangular libration point is illustrated in figure 8.8. Here the satellite is initially slightly displaced from the point of unstable equilibrium, and it soon begins to recede from the point. What began as a rigid circular motion about the center of mass is transformed into an irregular, chaotic revolution around body $A$.

### 8.4.3 The collinear libration points

The other three libration points are located on the straight line passing through the massive bodies. One of the points lies between the bodies. If their masses are equal, this point is just halfway between the bodies. (That is, it coincides with the center of mass of the system.) This case is discussed above (see figure 8.5). For a system with $m_B/m_A = 1/2$ (figure 8.9), the interior libration point is displaced from the center of mass toward the lighter body $B$ by 0.237 of the distance $AB$ between the bodies.

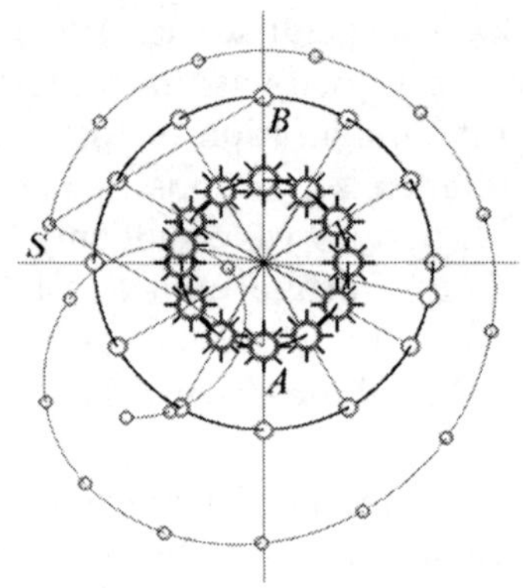


**Figure 8.8.** The instability of motion of the satellite in the vicinity of a triangular libration point ($m_B/m_A = 1/2$).

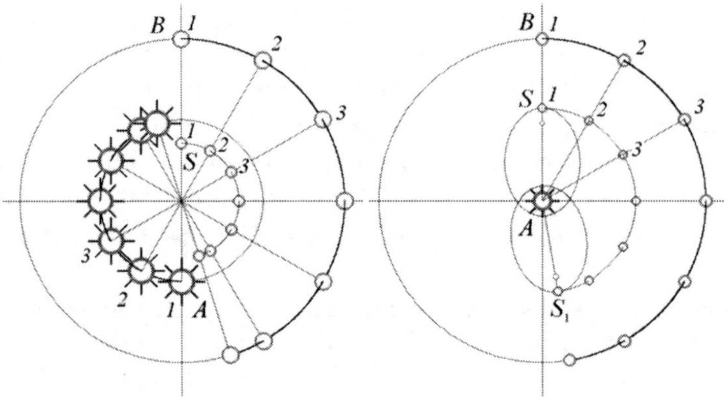


**Figure 8.9.** The motion of the satellite at the interior collinear libration point.

Its distance *SB* from the lighter body *B* equals approximately 0.43 *AB*, while the distance *SA* from the heavier body *A* is 0.57 *AB* (for detailed calculations of these distances see section 11.10). In this position the resulting force of gravitational attraction by the bodies *A* and *B* is directed toward the center of mass and its magnitude is just sufficient to provide the satellite *S* with the centripetal acceleration necessary for circular motion about the center of mass with the same angular velocity as that of the uniform rotation of the line *AB* joining the massive bodies. Thus, the rectilinear configuration of the system is preserved during the motion. The simultaneous positions of all the bodies are marked in figure 8.9 by equal numbers.

From the point of view of an observer on the heavier body *A* (see the right panel of figure 8.9), the lighter celestial body *B*, moving around *A* in a circular orbit, is continually eclipsed by satellite *S*, since the visible position of *S* always coincides with that of *B*. Similarly, an observer on *B* perceives the situation as a uniform revolution of *S* about himself in a circular orbit of radius 0.43 *AB*. This revolution visually coincides with the revolution of the celestial body *A* around *B*.

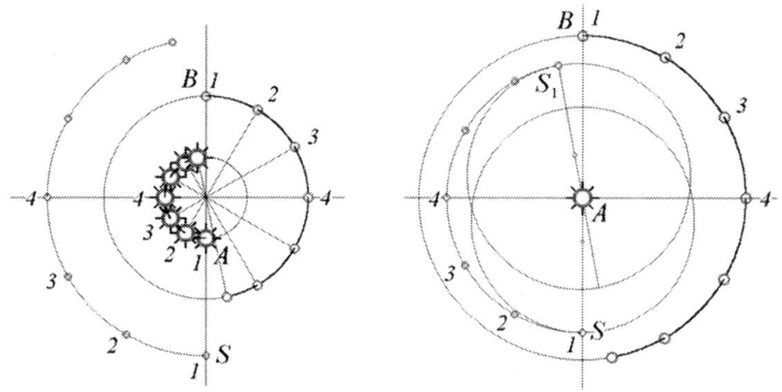


**Figure 8.10.** The motion of the satellite at the opposite collinear libration point.

The ellipses in the right panel of figure 8.9 show the osculating orbits that the satellite would trace around $A$ if $B$ were to suddenly vanish. (The first ellipse corresponds to the initial moment and the second ellipse to the moment when the satellite is at the point $S_1$.) Indeed, the circular velocity of the unperturbed orbital motion around $A$ is greater than the circular velocity of the actual motion, when the satellite is also subjected to the gravitational pull of the other body $B$. This additional pull reduces the centripetal acceleration of the satellite, and thus a smaller velocity is required for the circular motion.

Two collinear libration points lie outside the segment $AB$ joining the massive bodies. For the system with equal masses ($m_B = m_A$) these points are located symmetrically at a distance of 1.198 $AB$ from the center of mass, that is, at a distance of 0.698 $AB$ beyond either of the bodies. If $m_B < m_A$, one of the outer points is located closer to $B$. For $m_B/m_A = 1/2$, its distance from the center of mass equals 1.249 $AB$, so that this point of libration is separated from the lighter body $B$ by a distance of 0.582 $AB$. The opposite collinear libration point is located at a distance of 1.136 $AB$ from the center of mass, so that its distance from the heavier body $A$ equals 0.803 $AB$. The motion of the system with a satellite at this point in the inertial center-of-mass frame of reference is illustrated in the left panel of figure 8.10. The resulting gravitational pull of both $A$ and $B$ centripetally accelerates the satellite at this libration point so that it moves along its circular orbit with an angular velocity exactly equal to the angular velocity of the mutual revolution of $A$ and $B$.

The circular motion of the satellite at this collinear libration point in the reference frame of $A$ is shown in the right panel of figure 8.10. The osculating ellipses grazing the circular orbit of the satellite at the initial point $S$ and another point $S_1$ are the trajectories the satellite would follow if $B$ suddenly vanished at the moment the satellite is at these points. It may seem strange at first that the presence of $B$ (which produces an additional pull toward the center) reduces the centripetal acceleration of the satellite instead of increasing it. However, we remember that here the motion is referred to a non-inertial frame of reference associated with $A$. The gravitational field created by $B$ accelerates not only the satellite but also gives an even greater acceleration to $A$ and hence to the reference frame associated with $A$. Thus, the

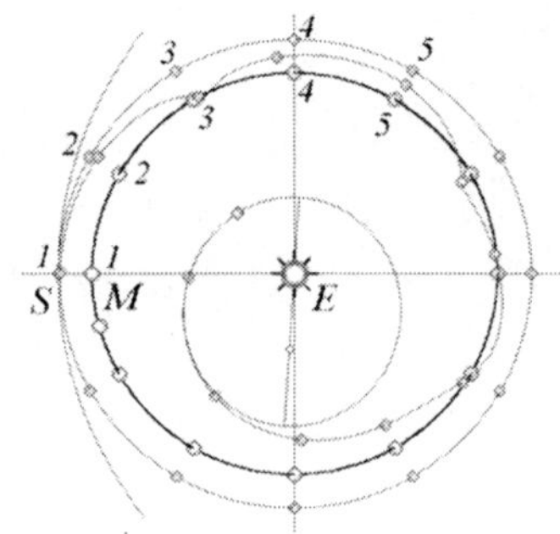


**Figure 8.11.** The instability of motion of the satellite at the exterior collinear libration point.

presence of the second massive body $B$ reduces the acceleration of the satellite in its motion relative to $A$.

For the Earth–Moon system, the distance of the interior libration point from the Moon equals approximately 58 000 km, or 0.15 of the mean distance $AB$ between the Earth and the Moon (384 400 km). The distance of the exterior point from the Moon equals 65 000 km, or 0.17 $AB$. The third collinear libration point lies on the opposite (with respect to the Moon) side of the Earth. Its distance from the Earth equals 380 600 km, or 0.993 $AB$.

The motion of a satellite in any of the collinear libration points (as well as the relative equilibrium in the rotating frame of reference) is unstable. The instability of the exterior libration point in the Earth–Moon system is illustrated in figure 8.11, which shows the motion of the system in the frame of reference associated with the Earth $E$. The initial position of the satellite $S$ is very close to the libration point. If the Moon $M$ were absent, the satellite would have moved in the gravitational field of the Earth along an ellipse, a part of which, grazing the actual circular orbit at the initial point $S$, is shown in figure 8.11. The additional gravitational pull of the Moon causes the satellite to move in a circle.

However, the satellite moves with the whole system in close proximity to the libration point (for the given initial displacement from the libration point) only during approximately one revolution. At the beginning of the second revolution it leaves the vicinity of the libration point and becomes a satellite of the Moon $M$. The orbital motion around the Moon is strongly perturbed by the Earth. After several revolutions about the Moon the gravitational field of the Earth tears the satellite away from the Moon and it becomes a satellite of the Earth. Its almost closed elliptical orbit is, in turn, perturbed by the Moon. The osculating unperturbed ellipse corresponding to this motion is also shown in figure 8.11.

Exact solutions to the three-body problem do not exist only for the circular motions considered above. When the two heavy bodies move around their center of mass in elliptical orbits, the third body of negligible mass placed at one of the five libration points can also move in a closed elliptical orbit provided its velocity has the value required for such motion. During such regular motions the distances between the bodies are subjected to periodic variations while the three bodies trace

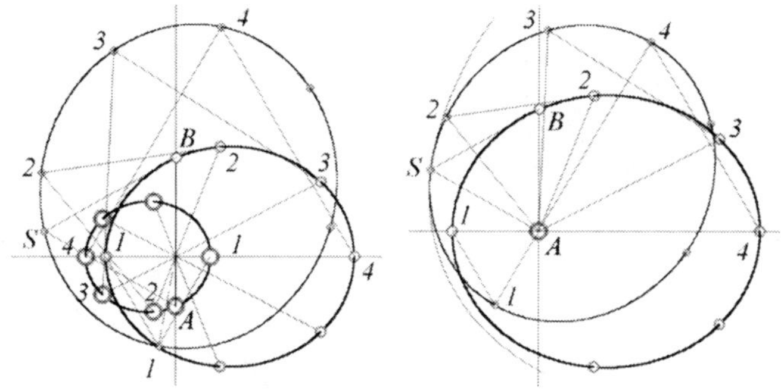


**Figure 8.12.** The periodic elliptic motions of bodies described by an exact particular solution of the restricted three-body problem with the light body at the triangular libration point.

homothetic ellipses with a common focus at the center of mass of the system. The program 'Planet with a Satellite' allows us to simulate these extraordinary motions.

Figure 8.12 shows the periodic motion described by such an exact solution with the satellite *S* at the apex of the equilateral triangle *ABS* whose base *AB* is the line joining the massive bodies *A* and *B* ($m_A/m_B = 2$). The left panel of the figure corresponds to the inertial center-of-mass reference frame in which all three bodies move in homothetic elliptical orbits. The right panel shows the motions of the satellite *S* and of body *B* in the frame of the heavier body *A*. The simultaneous positions of the bodies are marked by equal numbers.

The equilateral triangular configuration of the bodies is preserved during the motion; that is, the satellite remains at all times at the corresponding libration point. However, in contrast to the case of circular motion, here the triangle formed by the bodies rotates non-uniformly (together with line *AB* joining the bodies) and the lengths of its sides vary periodically during the motion (just as does the distance *AB* between the heavy bodies). The major axis of the ellipse traced by the satellite is at an angle with the major axes of the ellipses traced by the heavy bodies. The three bodies pass simultaneously through the corresponding points of their elliptical orbits (say, through the ends of the major axes). At points marked *1* in the figure, the bodies are at their shortest distances from the center of mass and their angular velocity (the same for all bodies) is greatest. At the remotest points *4* the angular velocity is smallest.

In the frame of body *A* (the right panel of figure 8.12) body *B* and the satellite *S* move in congruent ellipses around *A*. The major axis of the ellipse traced by *S* makes an angle of 60° with the major axis of the ellipse traced by *B*. If body *B* were suddenly to vanish, the satellite would leave its elliptical orbit and move after this moment along a larger osculating ellipse. A part of this osculating ellipse grazing the actual trajectory at point *S* is shown in the right panel of figure 8.12.

For $m_A/m_B = 2$, the motion described by this exact solution is unstable. After moving for a while in the proximity of the triangular libration point, the satellite

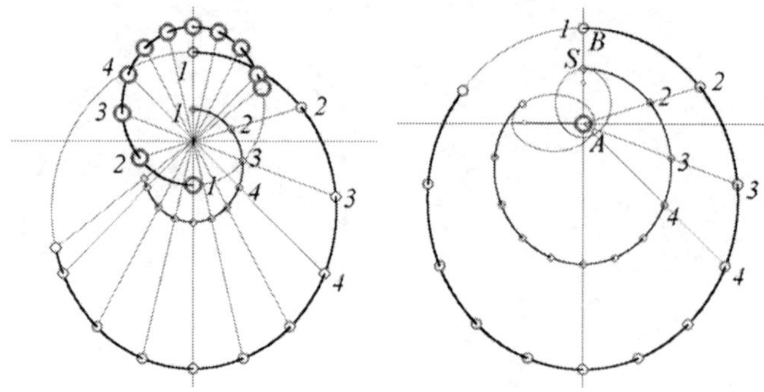


**Figure 8.13.** The periodic elliptic motions of the bodies described by an exact particular solution of the restricted three-body problem with the light body at the interior libration point.

irregularly orbits one of the bodies, its motion being strongly perturbed by the other body. Eventually it hits one of the bodies or is ejected from the system. (The duration of the regular portion of this motion depends on the precision with which the required initial conditions are entered in the simulation.)

The motion of the satellite at the interior Lagrange point of the two heavy bodies tracing elliptical orbits is illustrated in figure 8.13. In this motion the satellite $S$ remains between the heavy bodies, on the line joining them. This line rotates non-uniformly while the bodies move along the ellipses. The position of the satellite divides the line in a constant ratio and therefore the satellite traces an ellipse homothetic with those traced by the heavy bodies. The position of the interior Lagrange point between the heavy bodies depends on their masses in the same way as it does in the circular problem considered in detail in section 11.10. For example, in a system with $m_A/m_B = 2$ (figure 8.13), the interior libration point is displaced from the center of mass toward the lighter body $B$ through 0.237 of the distance $AB$ between the bodies. Its distance $SB$ from the lighter body $B$ equals 0.43 $AB$, while the distance $SA$ from the heavier body $A$ is 0.57 $AB$.

From the point of view of an observer on the heavier body $A$ (see the right panel of figure 8.13), the lighter celestial body $B$, moving around $A$ in an elliptical orbit, is continually eclipsed by satellite $S$, since the visible position of $S$ always coincides with that of $B$.

The small ellipses in the right panel of figure 8.13 show the osculating orbits that the satellite would trace around $A$ if $B$ were to vanish suddenly. The first ellipse corresponds to the initial moment and the second ellipse to the final moment of the simulation.

The motion of a satellite at any of the collinear Lagrange points is always unstable (whatever the ratio of the masses of the heavy bodies may be). That is, sooner or later its simple elliptical motion inevitably transforms into irregular, chaotic orbital motion around one of the bodies, and eventually ends with an ejection of the satellite from the system or with the satellite crashing against one of the heavy bodies.

## 8.5 A space flight over the back side of the Moon

The calculation of the trajectory for a space flight to the Moon is another example of the restricted three-body problem. As mentioned above, the problem does not have an exact analytic solution, but its approximate solution can be obtained by the method of joined conic sections, discussed in detail in section 10.11. The principal idea of the method is to ignore the influence of the Moon until the spacecraft crosses the boundary of the sphere of gravitational action of the Moon. In other words, we consider the motion in the geocentric frame of reference as an unperturbed Keplerian motion in the gravitational field of the Earth. After the spacecraft crosses the sphere of gravitational action of the Moon, we consider its motion in the non-inertial reference frame associated with the Moon (selenocentric frame) and assume that it is governed solely by the gravitational field of the Moon. On the boundary of the sphere we should transform the coordinates and velocity of the spacecraft from one frame of reference to the other. Thus the three-body problem is reduced to two two-body problems, for which the exact analytic solutions are available. This simple method proved to be very useful for a preliminary investigation of possible trajectories for space expeditions. Its reliability was proved by numerically integrating the differential equations of motion.

We can use the program 'Planet with a Satellite' to simulate a space flight to the Moon. Here we let the Earth be the greater of the two massive bodies (the 'star') and the Moon the smaller one (the 'planet'). We let the spacecraft remain initially near the Earth in a circular orbit, whose radius is, say, one tenth the distance between the Earth $E$ and the Moon $M$ (figure 8.14). In order that the spacecraft reaches the orbit of the Moon, we must increase its velocity. Using its on-board rocket engine, we direct its jet tangent to the circular orbit, opposite its orbital velocity. It then assumes a new elliptical orbit, whose perigee is at the initial point $S$. At this point the ellipse grazes the initial circular orbit. If we plan to obtain, say, a photograph of the back side of the Moon with the help of our automatic spacecraft, the apogee $A$ of the unperturbed ellipse should be somewhat greater than the radius of the Moon's orbit.

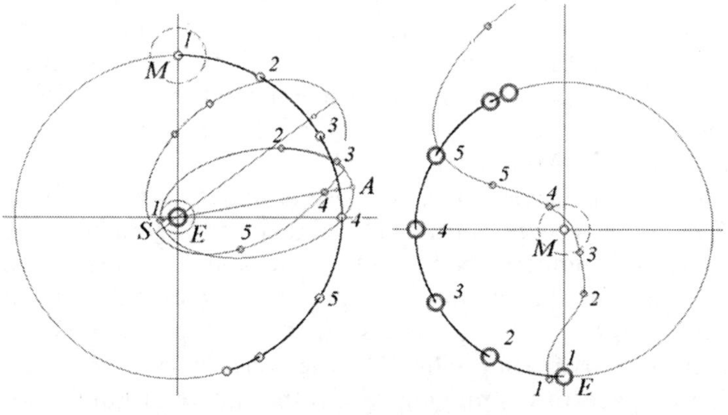


**Figure 8.14.** A space flight over the back side of the Moon.

A simple calculation based on the laws of energy and angular momentum conservation for motion in the Newtonian gravitational field of the Earth (see section 10.5 for details) shows that the required velocity at perigee (at the initial point *1* of the elliptical trajectory) must be approximately 4.26 times greater than the orbital velocity of the Moon (for the initial circular orbit whose radius equals 0.1 of the Earth–Moon distance). The spacecraft must reach the apogee of its elliptical orbit at the proper moment, namely simultaneously with the arrival of the Moon at the same position. This means that the radius to the starting point *1* must make a certain angle with the Earth–Moon line (about 100° opposite to the direction of the orbital motion).

The left panel of figure 8.14 shows the unperturbed elliptical orbit of the spacecraft in the geocentric frame of reference. The perigee is at the initial point *S* and the apogee *A* lies slightly beyond the Moon's orbit. The actual trajectory is obtained by numerically integrating the equations of motion that include the gravitational forces of both the Earth and the Moon. The simultaneous positions of the spacecraft and the Moon are marked by equal numbers. The geocentric velocity of the spacecraft near the apogee *A* of the elliptical orbit is smaller than the orbital velocity of the Moon. The Moon in its orbital motion overtakes the spacecraft and advances forward under it. The motion of the spacecraft over the back side of the Moon is retrograde relative the Moon's surface.

We see that for almost the entire trip toward the Moon's orbit, the actual trajectory nearly coincides with the unperturbed osculating Keplerian orbit. But close to the Moon the two trajectories noticeably diverge. The actual trajectory does not reach the apogee *A* of the osculating ellipse and abruptly bends down between the positions *3* and *4*. This divergence of the two trajectories is clearly caused by the gravitational pull of the Moon which is not taken into account in the osculating ellipse.

The right panel of figure 8.14 shows the motion of the spacecraft and the Earth in the selenocentric (fixed to the Moon) reference frame. We note again the retrograde character of motion of the spacecraft over the back side of the Moon.

The dashed circle around the Moon marks the boundary of the sphere of gravitational action of the Moon with respect to the Earth. The spacecraft enters the sphere (near point *3*) with a hyperbolic selenocentric velocity, and leaves the sphere (near point *4*) with a velocity that has the same magnitude but a different direction. The direction of the selenocentric velocity of the spacecraft is changed by the gravitational pull of the Moon.

After the spacecraft leaves the sphere of gravitational action of the Moon, its further motion in the geocentric frame of reference can again be treated with great accuracy as an almost unperturbed Keplerian motion. The actual trajectory of the spacecraft obtained in the simulation experiment practically coincides with the new osculating ellipse (the larger ellipse in the left panel of figure 8.14). The transition from one osculating ellipse to the other is caused by the Moon and occurs while the spacecraft is moving within the sphere of gravitational action of the Moon.

## 8.6 Lunar perturbations of a satellite's orbit

The gravitational field of the Moon accelerates both the Earth-orbiting satellite and the Earth itself. These accelerations are almost equal if the satellite is not far from the Earth. Therefore the lunar perturbations of the orbit of a satellite in its geocentric motion are caused not by the gravitational field of the Moon by itself, but rather by the non-uniformity of this field over distances that are equal approximately to the dimensions of the orbit. The difference between the actual acceleration of the satellite and the acceleration it would have in the absence of the Moon is called here the *perturbational acceleration.*

On the surface of the Earth, it is this non-uniformity of the gravitational field of the Moon (and to a lesser degree, of the Sun) that gives rise to the ocean tides. Since the gravitational field of a celestial body (say, the Moon) accelerates the Earth and the bodies on its surface by almost the same amount, only the *difference* in the force, exerted on terrestrial bodies located at different places, is important. This *differential* gravitational force is called the *tidal force.* Unlike the total gravitational force of the celestial body, the tidal force decreases as the cube, not the square, of the distance to the celestial body that causes the force (see section 11.13 for details).

There are several other situations in the Solar System where tidal forces come in to play. These forces give rise to the *Roche limit*—the closest distance of approach to a planet where a (natural) satellite can survive without being torn apart by tidal forces. In its simplest form, the Roche limit is defined as the distance from a massive body (planet) at which the self-gravitation of the smaller body (satellite) is just equalled by the tidal force due to the massive body.

A satellite orbiting the Earth is also perturbed by the tidal forces. The lunar gravitational perturbations are greater the larger the satellite's orbit. The direction of the perturbational acceleration depends on the position of the satellite relative to the Earth and Moon. If the satellite is directly between the Moon and Earth, the perturbational acceleration is directed away from the Earth toward the Moon because here the Moon-induced acceleration of the satellite is greater than the Moon-induced acceleration of the Earth. (However, its net acceleration is toward the Earth.) On the opposite side of the Earth, the perturbational acceleration is directed away from both the Moon and the Earth because here the Moon-induced acceleration of the satellite is smaller than the Moon-induced acceleration of the Earth. (But as before, the net acceleration is toward the Earth.)

When the radius vector of the satellite is at right angles to the Earth–Moon line, and hence the satellite is approximately at the same distance from the Moon as is the Earth, the perturbational acceleration is directed toward the Earth because the Moon-induced accelerations of the Earth and the satellite, though almost equal in magnitude, have slightly different directions. The vectorial difference between these accelerations is a small perturbational acceleration directed toward the Earth. The origin of the tidal forces is discussed in section 11.13 in greater detail.

The program 'Planet with a Satellite' allows us to simulate the lunar gravitational perturbations of a satellite's orbit. As in the preceding section, we let the Earth be the 'star' and the Moon be the 'planet'. Figure 8.15 shows the orbit of the satellite in

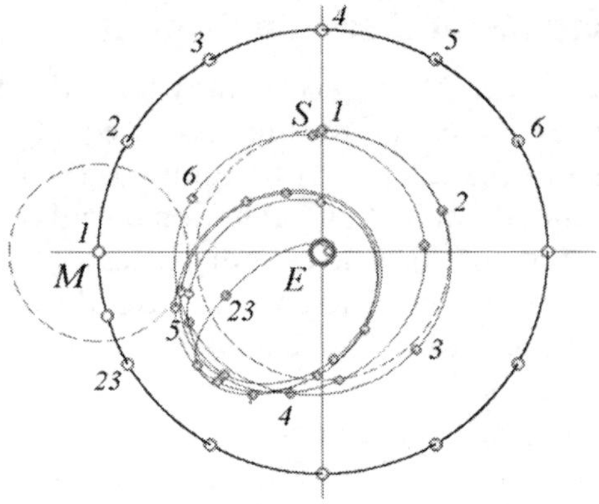


**Figure 8.15.** Lunar gravitational perturbations of a large orbit (initially circular) of an Earth-orbiting satellite.

the geocentric (Earth) reference frame. Initially the satellite *S* is launched to orbit the Earth *E* from point *1* in a circular path (shown by a dashed line in the figure) that almost reaches the sphere of gravitational action of the Moon *M*. The gravitational perturbations from the Moon cause the observed deviations of the actual trajectory from the osculating unperturbed orbit. When the satellite moves into the vicinity of points *2–3*, the Moon is on the opposite (left) side of the Earth so that the perturbational acceleration is directed to the right and the resulting geocentric acceleration is insufficient to keep the satellite in the original circular orbit. The actual orbit of the satellite deviates from the unperturbed circle. The osculating orbit becomes an ellipse.

During the further motion the perturbations primarily influence the apogee of the orbit, since there the perturbational acceleration is greater because of a greater distance from the Earth. Furthermore, at apogee the orbital speed is smallest so that here the perturbational acceleration is more influential on the orbit than at other points of the orbit. The influence of the perturbational acceleration at the apogee of the orbit is greater if the acceleration is directed parallel or antiparallel to the orbital velocity. That is, the effect is more pronounced when the large axis of the osculating ellipse is oriented neither along the Earth–Moon line nor perpendicular to it.

Lunar gravitational perturbations can cause significant changes in the satellite's orbit, including its destruction. The orbit obtained in the simulation shown in figure 8.15 eventually becomes so erratic from the Moon's perturbations that the satellite crashes into the Earth. To make the effect of gravitational perturbations more evident, the mass of the 'Moon' in this simulation is chosen to be several times greater than that of the real Moon.

## 8.7 A space voyage to a distant planet and back

There are many reasonable trajectories for a space voyage from the Earth to another planet in the Solar System. These trajectories differ in shape, duration of the voyage, energy expenditure (or, which is the same, the required initial velocity of the spacecraft), and navigational- and control-system requirements. These factors are not equivalent. Their role depends essentially on the purposes of the expedition. The needs of an automatic research vehicle and those of a manned interplanetary ship differ greatly.

In designing reasonable trajectories, we use the approximate method of joined conic sections. We divide the trajectory of the round-trip passive flight in the gravitational fields of the Earth, Sun and planet into several parts.

1. From the point at which the rocket engine is switched off (several hundreds of kilometers over the Earth) to the boundary of the sphere of gravitational action of the Earth.
2. From this boundary to the boundary of the sphere of gravitational action of the target planet, during which the ship is under the gravitational influence of only the Sun.
3. Within the sphere of gravitational action of the target planet.
4. Return path from the point of leaving the sphere of gravitational action of the planet to the boundary of the Earth's sphere of gravitational action.
5. Inside the sphere of gravitational action of the Earth from its boundary to the upper strata of the atmosphere.

It is assumed that during the first stage of the space voyage, the spacecraft is subjected only to the gravitation of the Earth; during the second, only to the gravitation of the Sun; during the third, only to the gravitation of the target planet; and so on. For each of the stages we use the corresponding frame of reference: the geocentric frame for the first stage, the heliocentric frame for the second, the reference frame of the target planet (the planetocentric frame) for the third stage and again the heliocentric frame for the fourth stage.

At the boundaries of the regions we make a transition in our calculations from one reference frame to another by adding vectorially the corresponding velocities. (That is, to obtain the velocity of the spacecraft relative to the frame it is entering, we add vectorially the velocity of the spacecraft with respect to the reference frame it is leaving to the velocity of that frame relative to the frame it is entering.) Of course, in outer space there are no actual impenetrable boundaries for gravitational fields. The method of joined conic sections is an approximation only and its results should be confirmed by a direct numerical integration of the equations of motion.

The massive outer planets (Jupiter through Neptune) have almost circular orbits of large radii: 5.2 astronomical units (mean Earth–Sun distances) for Jupiter, 9.5 for Saturn, 19 for Uranus and 30 for Neptune. Their spheres of gravitational action have large sizes (for Jupiter it is more than fifty times greater than for the Earth) because of their large masses and great distances from the Sun. With the help of the program 'Planet with a Satellite' we can simulate a space voyage to some outer planet (Jupiter or Saturn) of the Solar System, assuming in the program the greater massive body (the 'star') to be the Sun, the smaller one to be the target planet and the satellite (the zero-massed body) to be the space vehicle.

We let the radius of the orbit of the target planet be, for example, ten times the radius of the Earth's orbit. Against the background of such a large distance, we can neglect the dimensions of the Earth's sphere of gravitational action (its radius is

150 times smaller than the mean Sun–Earth distance). Therefore we can simulate the second and third stages of the space voyage to the distant massive planet, and the way back to the vicinity of the Earth, as a three-body problem. For this purpose we should enter the heliocentric position and velocity of the spacecraft at the boundary of the gravitational sphere of the Earth (for the moment when the spacecraft leaves the sphere) as the initial conditions for the simulation.

The initial distance of the spacecraft from the Sun can be assumed to be equal to the radius of the Earth's orbit. To estimate the required initial velocity, we can neglect the gravitational attraction of the spacecraft by the target planet. The aphelion of an elliptical heliocentric orbit of the spacecraft must reach the circular orbit of the planet. The minimal necessary characteristic velocity of the spacecraft is required for the semielliptic (Hohman's) transition from the Earth's orbit. For such a transition the additional velocity of the spacecraft must be directed forward—tangentially to the orbital velocity of the Earth.

A calculation based on the laws of energy and angular momentum conservation for motion in the Newtonian gravitational field of the Sun (see section 10.5 for details) shows that the spacecraft should leave the sphere of gravitational action of the Earth with a heliocentric velocity that is approximately 4.26 times greater than the orbital velocity of the target planet whose orbit is ten times greater than that of the Earth.

When the outer planet is to be the target, a very important consideration is timing. The spacecraft must reach the aphelion of its heliocentric trajectory simultaneously with the target planet, just when the planet in its circular orbital motion approaches this aphelion. Therefore the angular position of the point to start en route from the Earth's orbit must be determined properly. We can do so by calculating the duration of motion of the spacecraft along the semielliptic trajectory with the help of Kepler's third law.

Taking into account the distance that the target planet covers during this time, we find that the starting point of the spacecraft must have an angular position of approximately 100° behind the angular position of the target planet at the starting time. Entering these initial values in the simulation experiment, we can expect that the spacecraft will reach the sphere of gravitational action of the target planet. Then, by trial and error, we can find those precise initial conditions that produce a desirable trajectory.

Near the aphelion of its heliocentric trajectory, the spacecraft, though traveling in the same direction as the planet, moves much more slowly (relative to the heliocentric reference frame). The planet overtakes the space vehicle and the craft enters the planet's sphere of gravitational action from the planet's forward side. Relative to the planetocentric frame, the velocity with which the spacecraft enters the sphere of gravitational action of the planet is greater than the parabolic velocity that corresponds to the gravitational field of the planet at the boundary of the sphere. (This relationship applies to all planets of the Solar System and to all possible transition trajectories.) Therefore the planetocentric trajectory of the space vehicle within the sphere of gravitational action of the target planet is always a hyperbola.

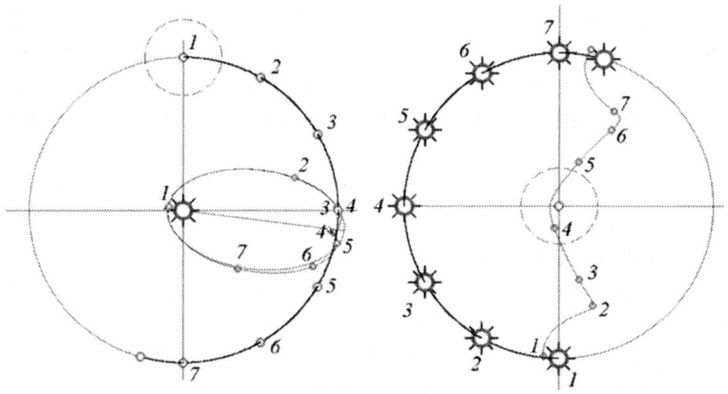


**Figure 8.16.** A space voyage to a distant planet and back.

This means that after it enters the sphere, the space vehicle must inevitably leave the sphere unless it strikes the planet or at least its atmosphere. In the planetocentric frame of reference, the exiting velocity has the same magnitude as does the entering velocity, but its direction is along the other asymptote of the hyperbola. To find the heliocentric exiting velocity, we should add the planetocentric velocity vectorially to the orbital velocity of the planet. The resulting exiting velocity generally differs from the entering one both in direction and magnitude. This velocity determines the further heliocentric motion of the space vehicle. If the spacecraft is to return to the vicinity of the Earth, we should plan to have this new heliocentric trajectory graze or intersect the Earth's orbit.

Figure 8.16 shows the trajectory of the space voyage both in the heliocentric frame of reference and the frame associated with the target planet. To make the features of the motion within the sphere of gravitational action more evident, we have chosen an exaggerated value for the mass of the planet (3% of the mass of the Sun) and, consequently, of the radius of the sphere, whose boundaries are shown by dashed circles in the figure. The simultaneous positions of the spacecraft, the target planet and the Sun are marked by equal numbers. The spacecraft leaves the Earth's sphere of gravitational action with a heliocentric velocity that is parallel to the Earth's orbital velocity and 4.21 times the orbital velocity of the target planet (approximately 1.33 units of the Earth's orbital velocity) at point *1*, whose angular position is approximately 83° behind that of the target planet.

The unperturbed elliptical heliocentric orbit of the spacecraft is shown in the left panel of figure 8.16. The spacecraft follows the unperturbed ellipse almost exactly up to position *3*, where it reaches the planet's orbit. We note that the planet at that moment is rather far behind the point. The further heliocentric motion of the spacecraft between positions *3–4* is considerably distorted by the gravitation of the planet. The planet gradually overtakes the slowly moving spacecraft, and the spacecraft appears inside the sphere of the gravitational action of the planet.

Inside the sphere the planetocentric motion of the spacecraft is governed mainly by the gravitation of the planet. The hyperbolic part of the planetocentric trajectory inside the dashed circle is clearly seen in the right panel of figure 8.16. The tiny loop in the heliocentric trajectory between positions *4* and *5* (near the distant focus of the unperturbed ellipse in the left side of the figure) is produced by addition of the hyperbolic planetocentric motion and the uniform circular orbital motion of the planet.

To ensure that the spacecraft finds a way back to the neighborhood of the Earth after passing near the planet, we can choose the initial conditions of departure (for example, by trial and error) in a way to make the unperturbed heliocentric ellipse for the return to the Earth almost equal (congruent) to the ellipse of the outward bound motion. In our simulation these ellipses (the ellipse for the return is not shown in the figure) differ slightly only in the orientation of their major axes. The osculating ellipse for the return has the same parameters (major axis and eccentricity) as that for the outward voyage if the two points at which the trajectory enters the sphere of gravitational action and leaves it lie symmetrically at equal distances from the Sun, and if the vectors of the heliocentric velocities at these two points are equal in magnitude and make equal angles with the line joining the points. This arrangement occurs if the axis of symmetry of the planetocentric hyperbolic trajectory is orthogonal to the line joining the points in the heliocentric frame.

Since the length of the major axis and the distance of the perihelion from the Sun are the same for the ellipse of the return trip as they are for that of the outward voyage, the trajectory returns the spacecraft to the Earth's orbit in a time equal to that of the outward voyage. But, to place the spacecraft into the circular orbit of the Earth, an additional rocket thrust at the perihelion of the elliptical orbit is required. The resulting change in velocity must be of the same magnitude as the initial one, but opposite in direction.

To economize on rocket fuel, we should be concerned about the duration of the designed expedition: if the spacecraft arrives at the perihelion of its elliptical return path at the instant when the Earth in its orbital motion is just at this point, the Earth's atmosphere can be used to quench the excess velocity of the spacecraft.

To provide an entrance into the atmosphere at a small angle, we should plan, in a real voyage, to use additional rocket impulses to correct the trajectory. And certainly all the calculations of the trajectory for this last stage of the voyage (inside the Earth's sphere of gravitational action) must take into account the gravitation of the Earth.

## 8.8 Comets—interplanetary vagabonds

In spite of their small masses compared to the masses of the planets, comets are visible even to the naked eye when they approach the Sun. The nuclei of most comets are thought to be a kind of dirty snowball no more than a few kilometers in diameter. The comet's nucleus is surrounded by a coma—a nebulous cloud of gas and dust. Passing near the Sun, some of the comets acquire a tail that is always directed away from the Sun. When the Sun's radiation heats the nucleus, a tail of evaporated (sublimated) gases is formed and stretches through great distances under the pressure of light and the solar wind. The length of the comet's tail can exceed the

Sun–Earth distance. The long bright tail in combination with its small mass prompts us to call such a strange heavenly vagabond a 'visible nothing'.

As members of the Solar System, these bodies travel around the Sun in extremely eccentric orbits. Some of them have very long periods, exceeding 100 000 years. The elongated elliptical orbits of such long-period comets are almost indistinguishable from parabolas, especially because we are able to observe only the small portions of their orbits that lie in the vicinity of the Sun. These comets appear unexpectedly, in contrast to short-period comets with orbital periods of less than 150 years.

The most famous short-period comet was recognized as such in 1705 by Edmund Halley (1656–1742). He established that the comets observed in the years 1531, 1607 and 1682 were in fact a single heavenly body that periodically returns to the Sun approximately every 76 years. Its last visit was in 1986. The aphelion of its elliptical orbit (of retrograde revolution) is beyond the orbit of Neptune, the remotest giant planet of the Jovian group. Currently we are aware of many short-period comets with periods from three to ten years.

The masses of comets are too small to influence the motion of the planets even when the comet passes very close to a planet. On the other hand, massive planets such as Jupiter do change the orbits of comets significantly. Depending on the approach of the comet to the planet, and on the velocity of their relative motion, a 'gravitational slingshot' or 'gravity assist maneuver' with the planet can either increase or diminish the eccentricity and size of the comet's orbit. If the orbit is increased, the period is also increased. The comet may even be transferred to an open hyperbolic orbit and ejected from the Solar System. Conversely, if the eccentricity is decreased, a long-period comet may be trapped in a small orbit with a short period.

Like the major planets of the Solar System, most short-period comets have orbits with rather small inclinations to the plane of the Earth's orbit, that is, to the ecliptic. We can simulate the planetary perturbations of such a comet with the help of the program 'Planet with a Satellite', which deals with the motion of three bodies whose orbits lie in a common plane. Here the satellite (a body of a negligible mass) plays the role of a comet.

In order to make the effect of planetary gravitational perturbations of the comet easily observable in the simulation, we choose the initial eccentric orbit of the comet with a perihelion distance smaller than the radius of the planet's orbit, so that the orbits of the comet and the planet intersect.

Figure 8.17 illustrates such a simulation. The initial position of the comet $C$ is near the Sun $S$ on the opposite side of the planet $P$ (say, Jupiter), which moves around the Sun (counterclockwise) in a circular orbit. The dashed circle around the planet $P$ shows the sphere of its gravitational action relative to the Sun. The comet makes a revolution along its initial elongated elliptical orbit during approximately seven revolutions of the planet, that is, during seven Jovian years. The small circles with numbers show the position of the comet at the moments the planet completes the corresponding revolution and returns to $P$.

If the comet passes near the Sun (through the perigee of its orbit) when the planet is far from the point (say, on the opposite side of the Sun), the orbit of the comet is

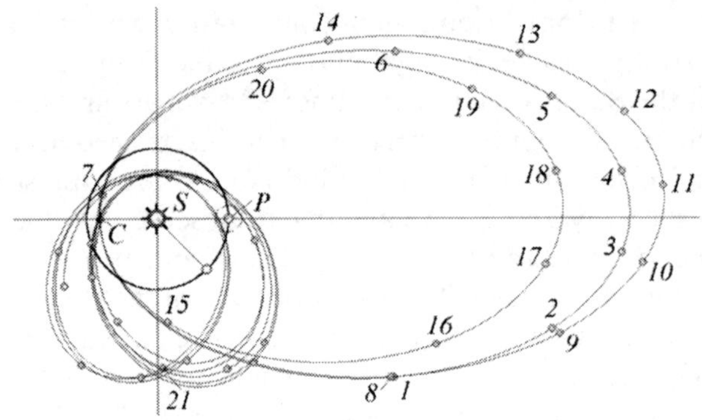


**Figure 8.17.** The perturbations of a short-period comet caused by a massive planet. The circles with numbers show the position of the comet after each revolution of the planet.

only slightly perturbed by the planet. Such a situation occurs after the comet's first revolution, which lasts almost a whole number of Jovian years (position *7* in figure 8.17, the planet being at point *P* at the moment). Hence in the second revolution the comet traces an orbit that differs only slightly from the original one. However, the period of revolution is increased, and so at the next passage of the comet through perihelion (between positions *14* and *15*) the planet is closer and perturbs the comet's orbit more greatly. These rather moderate perturbations by a massive planet distort the comet's elongated heliocentric orbit near its perihelion and cause irregular variations in the period of revolution. Most of the short-period comets exhibit such variations. For example, the period of revolution of Halley's comet varies between 74 and 79 years.

Sooner or later the motions in the intersecting orbits with non-commensurate periods bring the comet into close proximity to the planet. In the simulation shown in figure 8.17, this rendezvous occurs after the third revolution of the comet, between positions *20* and *21*. The comet enters the sphere of gravitational action of the planet with a hyperbolic velocity relative to the planet. Considering the planetocentric motion of the comet (that is, the motion of the comet referred to the frame associated with the planet), within the sphere of gravitational action of the planet, we can neglect the attraction of the comet by the Sun. Tracing a portion of the hyperbola (in the planet's reference frame), the comet leaves the sphere of gravitational action (provided the hyperbola does not intersect the surface of the planet) with a velocity of the same magnitude (relative to the planet), but in a different direction. That is, the comet cannot be captured by the planet to become a satellite of the planet.

The result of this encounter of the comet with the planet reminds us of a perfectly elastic collision, in which the *relative velocity* changes only in direction. We can use this analogy with elastic collisions because the comet crosses the sphere of gravitational action during an interval that is small compared to the period of its revolution around the Sun. The resulting relatively rapid change in direction and magnitude of the heliocentric velocity of the comet produced by the encounter with the planet can be treated as a gravitational slingshot or gravity assist maneuver that causes a considerable variation in the heliocentric orbit of the comet.

In several projects of space expeditions, similar gravitational interactions with planets were used to deliberately change the velocity and trajectory of an automatic space probe by taking a small amount of the planet's orbital energy. For example, the Voyager mission was designed by NASA to take advantage of an unusual alignment of the outer planets during the 1970s and 1980s. The configuration of the Jovian planets in the late 1970s (alignment of Jupiter, Saturn, Uranus and Neptune) allowed Voyager to increase its heliocentric velocity with the help of a series of gravitational slingshots with these outer planets and thus to visit and explore at short distances several of these planets in one expedition lasting about 12 years. The Voyager Grand Tour missions profoundly extended our knowledge about the remote planets of the Solar System and their numerous satellites. A similar alignment of the Jovian planets will not occur again until the middle of the 22nd century.

In the simulation shown in figure 8.17, the comet experiences a strong gravitational interaction with the planet after three revolutions along its initial, almost elliptical orbit, which is rather moderately perturbed by the planet. Because of the interaction, the comet is captured in a small, short-period heliocentric orbit. This orbit also cannot exist for long since it intersects the planet's orbit. After five revolutions along the new orbit, the comet again meets the planet and this second gravitational interaction primarily influences the orientation of the major axis of the comet's orbit.

It is clear that the fate of the comet is extremely sensitive to the initial conditions, one more example of dynamical chaos: a very small initial difference may produce an enormous change in the future motion of the system. For example, the gravitational influence of the planet can so increase the heliocentric velocity of the comet that it is ejected from the system. Or the gravitational interaction with the planet during their approach can change the orbit of the comet so that after several revolutions it strikes the planet. With the simulation program, we can verify that a small variation in the system parameters or initial conditions may cause a quite different long-term consequence.

## 8.9 A double star with a planet

The simulation program 'Double Star with a Planet' is similar to the program 'Planet with a Satellite' discussed above. Both programs deal with the restricted three-body problem concerning the motion of a body of negligible mass under the gravitational pull of two massive bodies orbiting one another in circles or Keplerian ellipses. As already noted, the difference between these situations (the satellite of a star-orbiting planet or the planet in a double-star system) is quantitative rather than qualitative. Therefore the simulation of the satellite of a massive planet that orbits a star can also be regarded as the simulation of the motion of an inner planet, one that orbits one of the stars in the binary system (see figures 8.1 and 8.2). However, the method of setting the initial conditions and parameters of the system used in the program 'Double Star with a Planet' is more convenient for the simulation of the motion of an outer planet, one that orbits both stars.

To simulate this motion, we first enter the parameters of the system and the initial conditions in the panel 'Settings', opened by clicking the corresponding menu item. For the two stars, we then enter the ratio of their masses and the initial velocity that determines their relative motion. This velocity is directed transversely (that is, perpendicularly to the line joining the stars). Therefore we need enter only its magnitude, expressed in units of the circular velocity.

Next we enter the initial position and velocity of the planet. To do so, we can choose one of the following frames of reference: that associated with either of the stars, or the inertial reference frame associated with the center of mass of the system. The center-of-mass reference frame is convenient for the simulation of the motion of an outer planet. The choice of the reference frame is made by clicking the appropriate option button. The initial position of the planet is defined in terms of its distance from the chosen star (or from the center of mass if that is the chosen frame) measured in units of the initial distance between the stars, and the angle which the radius vector of the planet drawn from the star (or from the center of mass) makes with the straight line joining the stars.

The initial velocity of the planet is indicated in the same way. We first enter its magnitude. If we use the reference frame of one of the stars, the magnitude of the initial velocity must be expressed in units of the unperturbed circular velocity of the planet for its motion around the corresponding star (the velocity for which the satellite moves in a circular orbit about the star in the absence of the gravitational influence of the other star). These units of velocity are convenient for the simulation of the motion of an inner planet. If the center-of-mass reference frame is chosen, the magnitude of the initial velocity must be expressed in units of the circular velocity with which the planet would move around the center of mass under the assumption that the mass of both stars is concentrated there. Then we enter the angle between the initial velocity and the initial radius vector of the planet (drawn from the corresponding star or from the center of mass).

The behavior of a planet in the double-star system can be very complicated. Figure 8.18 shows an example of the irregular looping trajectory traced by a planet $P$ in the frame of reference of one of the stars. Initially the planet chaotically orbits the smaller star $B$ that moves around the greater star $A$ in an elliptical Keplerian orbit. Then star $A$ captures the planet, which then revolves around $A$ for a while in a small, almost elliptical orbit. After one revolution about $A$, the star $B$ recaptures the planet and during the next revolution of $B$ around $A$ the planet revolves around $B$ in a small orbit, following $B$ in its orbital motion around $A$.

Transitions from orbiting one of the stars to orbiting the other star and back occur several times. Such an irregular, chaotic motion of the planet ends in a collision with one of the stars (position $C$ in figure 8.18).

The long-term behavior of the system is very sensitive to the initial conditions. With slightly different initial conditions, the planet may strike the other star or be ejected from the system.

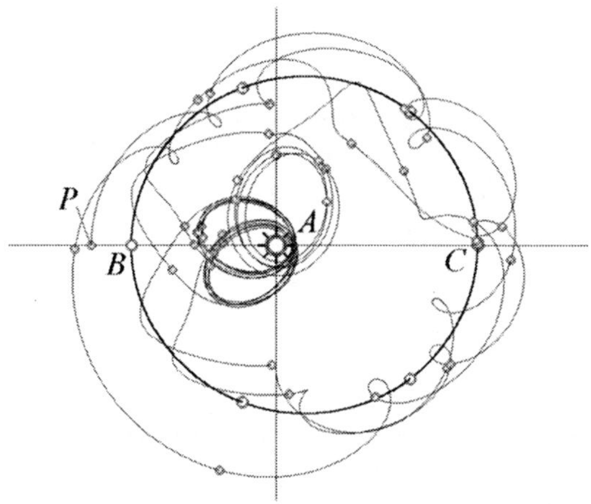


**Figure 8.18.** An example of the irregular planetary motion in a binary star system.

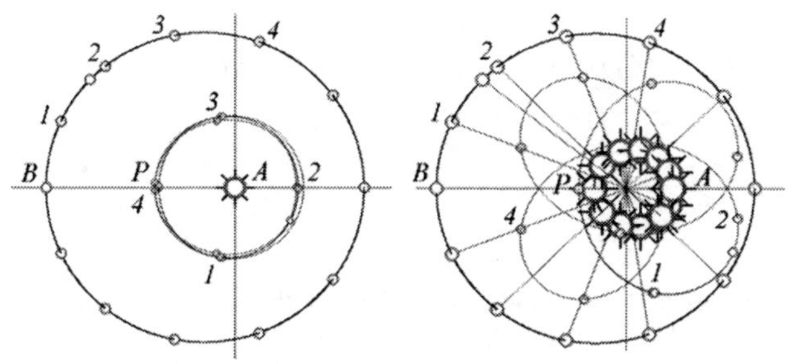


**Figure 8.19.** A periodic inner planet in the binary star system.

However, it is possible for a planet in a double star system to move regularly in a stationary orbit. Figures 8.1 and 8.2, show examples of such motions. A small, almost circular orbit around one of the stars is only slightly perturbed by the other star, and this motion can continue indefinitely. This situation is similar to the familiar example of a Moon orbiting a planet orbiting the Sun.

It may seem surprising that it is possible for a planet in a double-star system to move in a large, regular, periodic orbit in spite of the perturbations by the other star. Figure 8.19 shows an example of an inner planet $P$ moving (counterclockwise) around the heavier star $A$ in an orbit whose dimensions are only about a half of the orbit of the stars' (clockwise) relative motion. In the reference frame of star $A$ (left panel of figure 8.19), the planet makes exactly three revolutions around star $A$ during one period of mutual revolution of the stars. Although perturbed by the other star, the planet's orbit nevertheless is very nearly closed after three revolutions. When the smaller star $B$ completes a revolution, the planet arrives at the initial spatial point $P$ with the same velocity it had initially. The initial state of the system is almost exactly reproduced and the motion is nearly periodic.

In the inertial center-of-mass frame (right panel of figure 8.19), this regular motion of the three-body system looks even more complicated. The trajectory of the

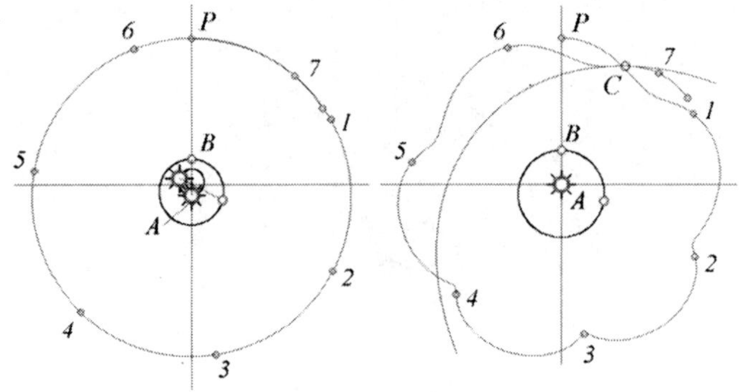


**Figure 8.20.** A stationary outer planet in the binary star system.

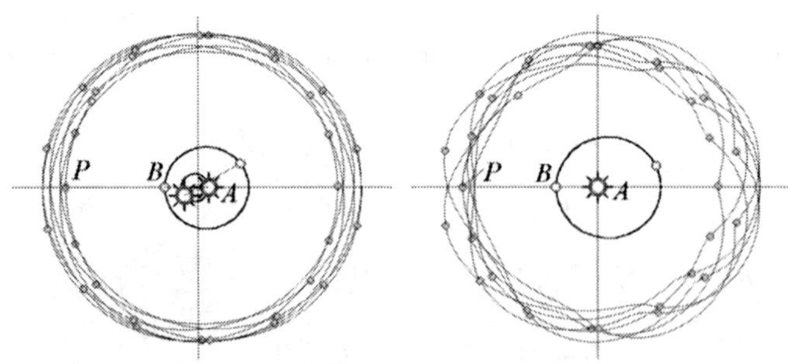


**Figure 8.21.** A periodic outer planet in the binary star system.

planet is a beautiful four-petalled closed curve, traced once during exactly one period of the stars' mutual revolution.

A planet can move steadily around a binary system in an outer orbit that encompasses both stars. If the orbit of the planet is large enough compared to the star–star distance, in the center-of-mass reference frame it is almost elliptical and practically closed. An example of such an orbit encircling the stars that trace elliptical orbits around the center of mass is shown in figure 8.20. Although the period of the planet's revolution is generally non-commensurate with the star's period, the motion of the planet is stationary and lasts indefinitely.

The right panel of figure 8.20 shows this motion in the reference frame of the heavier star $A$. The wavy shape of the planet's trajectory in this frame is explained by the periodic motion of star $A$ relative to the center of mass. The figure also shows a portion of the osculating elliptical orbit for point $C$ of the actual trajectory. The planet would trace this ellipse with the focus at $A$ if the second star $B$ were suddenly to vanish when the planet is at point $C$.

For outer planets, periodic motions along closed orbits are also possible. Figure 8.21 shows a surprising example of such a motion. Initially the stars $A$ and $B$ are found at the pericenters of the elliptical orbits they trace around the center of mass. The initial position of the planet $P$ is on the same line with the stars and the initial velocity of the planet is perpendicular to this line. In this example the outer orbit of the planet encircling both stars cannot be regarded as large compared with the star–star distance.

The small circles show the positions of the planet after each revolution of the stars, when the stars are again in their initial configuration (*A* and *B* in figure 8.21). The individual loops of the planet's trajectory are not closed since under the gravitational attraction toward the two moving centers the planet does not trace a Keplerian ellipse.

However, after seven revolutions the planet arrives at the initial point *P* with the same velocity it had initially and the trajectory of the planet is closed. During the time taken by the planet to make these seven revolutions in an outer orbit, the stars make thirty mutual revolutions, after which the initial state of the whole system is almost exactly reproduced! The right panel of figure 8.21 shows the wavy closed trajectory of the planet in the reference frame of the heavier star *A*.

## Questions and problems

1. **On parallel and opposite courses.** In which case are mutual gravitational perturbations of two planets greater: if the planets move around the star along almost circular orbits in the same direction, or if they move along the same orbits in opposite directions?
2. **Satellites with direct and retrograde revolutions.** Explain, in which case the satellite of a planet will experience greater perturbations from the star—if the revolution of the satellite around the planet occurs in the same sense (the same direction) as the revolution of the planet around the star, or in the opposite sense (the latter case is referred to as a retrograde revolution). Verify your answer by the simulation experiment.
3. **Exterior and inferior planets.** Explain why in the planetary system of a binary star exterior planets can have stable orbits only if the dimensions of such orbits are much greater than the dimensions of the binary star orbits, and, otherwise, inferior planets can have stable orbits only if their dimensions are much smaller than the dimensions of the binary star orbits. Why are stable motions of planets in orbits whose dimensions are comparable with those of the binary star components impossible?

**IOP** Publishing

Motions of Celestial Bodies
Computer simulations
**Eugene Butikov**

# Chapter 9

# Many-body systems in celestial mechanics

The simulation program 'Planetary System' allows us to investigate a model of the Solar System, or to create an imaginary planetary system of our own—complete with the star, planets, moons, comets, asteroids and satellites, and to explore their orbital motion governed by the gravitational forces. The simulation is based on the numerical integration of the differential equations of motion for the many-body system. The forces of gravitational attraction between all pairs of bodies constituting the system are taken into account. The masses of all the bodies, their initial positions and the distances between them, and their initial velocities can be chosen arbitrarily. If we enter in the program values of the same order of magnitude for the masses of the 'planets' as for the star, the simulation will refer to a multiple star rather than to a planetary system. The only restriction is that all the bodies and their velocities must lie in the same plane. This limitation is related solely to the difficulties of representation and visual perception of motion of a three-dimensional system on a two-dimensional computer screen.

The simulated motion can be displayed either in the 'heliocentric' reference frame (associated with the star), in the inertial frame associated with the center of mass of the system, or in the 'geocentric' reference frame (associated with one of the planets). The motion can be also displayed on the screen in any two of these frames simultaneously.

## 9.1 Planetary system—a many-body problem

When the program 'Planetary System' simulates the system of three or more bodies, their motion can be very complicated. The system can evolve into a new configuration that does not at all resemble the initial configuration. In the general case, the initial configuration is never reproduced. However, this evolution is reversible since the equations of motion of a conservative system are symmetric with respect to the time reversal. The program allows us to illustrate this reversibility of motion in the simulation experiment. There is an item 'Reverse Velocities' under the menu

doi:10.1088/978-0-750-31100-7ch9  

'Options'. If we click it, the program instantly reverses the directions of the velocities of all the bodies, and we can observe how they move backward along the same trajectories toward the initial configuration. Even if we start from a symmetric configuration of the bodies, the system loses this symmetry sooner or later during the motion. However, this evolution of the system toward less symmetric configurations cannot be regarded as its intrinsic property: the experiments with reversing velocities show clearly that the laws of motion also allow the evolution toward more symmetric configurations. Certainly, we may regard such cases of evolution as very rare and improbable because they require quite specific initial conditions.

The reversibility of motion is violated in cases of collisions of the bodies because such events are treated by the program as completely inelastic.

The program allows us to change the scale in which the motion is displayed. We can do this by clicking the menu items 'Zoom In' or 'Zoom Out'. Each time we do this, the scale is, respectively, increased or reduced by a factor of 1.25. We can also select some region of the window which is of most interest and enlarge this selected region to fit the whole window. To do this, we draw a rectangle with the mouse in the same way as we do when selecting part of the window in such popular graphics editors as Microsoft Paint. After we release the left button of the mouse, the boundaries of the rectangle may automatically change a bit to display a region whose sides are proportional to the corresponding sides of the window. We can move the rectangle as a whole into a new position, if necessary. (To deny the selection, we click somewhere outside it, or click the right button anywhere.) To expand the selected region over the whole window, we choose 'Zoom In' from the menu, or simply double-click inside the selection. If the motion is displayed simultaneously in two reference frames, we can make the selection for zoom either in one of the windows, or in both windows at once.

To enter the parameters of the modeled system, we open a special panel by clicking the menu item 'Input'. The panel shows the initial configuration of the planets and the list in which their masses, distances from the star, velocities and radii are indicated.

If we wish to create a new model, we click the button 'Clear' to remove all the old planets. Then we enter the mass of a planet (in units of the star's mass), its initial distance (in astronomical units, i.e. in units of the mean Sun–Earth distance), its angular position (in degrees), its initial velocity in units of the unperturbed circular velocity (i.e. velocity in the circular orbit if all other planets were absent), the direction of the initial velocity (the angle it makes with the radius vector in degrees) and the radius of the planet in units of the star's radius. These parameters can be entered into the corresponding boxes in any sequence.

When all the parameters are chosen, we click the button 'Add', and the planet is added to the list. Then we repeat the procedure for the second planet, the third, and so on. When we add a planet to the system, the new configuration of the planets with their unperturbed theoretical orbits appears in the window.

To choose the reference frame in which the motion is to be displayed, we selected the corresponding check-box below the list. We can choose the heliocentric frame (that associated with the star), the inertial center-of-mass frame, or the 'geocentric' frame, that is, the frame associated with one of the planets (namely, the planet that is the first one in the list). We can associate the 'geocentric' frame with any of the planets by moving the desired planet to the top of the list. To display the motion simultaneously in two reference frames, we check both corresponding boxes. To begin the simulation, we click the 'Ok' button. We can change the chosen reference frame at any time during the simulation by picking a desired frame from the menu 'View'.

To modify the modeled system, we can remove any planet from the system in the panel 'Input' by selecting the planet in the list and clicking the 'Remove' button. Then we can add a new planet (or several planets) to the list by using the procedure described above. If we are going to modify one parameter (or several parameters) of a definite planet, we double-click it in the list. Then the boxes used for entering the data will display the values of parameters that correspond to this planet. (When the panel is opened, these boxes show the values that correspond to the last planet in the list.) After changing the values in the boxes, we click the button 'Add', and a planet with the new parameters appears at the end of the list.

However, if we are only going to modify some parameters of the planet (but not to add a new planet to the system), we should remove the planet with unmodified parameters from the list. And if we do not modify the initial position of the planet, we should remove the old planet before we click the 'Add' button because it is not possible in this simulation to position two planets at the same spatial point.

The menu item 'Examples' opens a panel that displays a set of predefined examples that the program offers. Option buttons 'Basic set' and 'Extended set' allow us to switch between the two sets of examples. When we select an example from the list, its brief description appears in the text-box below. To start an example, we double-click it in the list. To display the parameters of the system before starting an example, we click the 'Ok' button, after which the panel 'Input' is opened with the list of the planets and their parameters. Clicking the 'Ok' button of this panel, we launch the simulation.

To create new examples of our own, we choose the option button 'Custom set' in the panel 'Examples'. The menu item 'Edit' is then enabled and allows us to modify an existing set of examples by removing some of its items and adding new ones, or to create new sets and store them in files. Choosing 'Edit name and comment', we can change the title and/or description of any example without changing the parameters of the system. The items 'Remove example', 'Move example up' and 'Move example down' allow us to organize the set.

To add a new example to a set, we first construct a planetary system (with the help of the panel 'Input', see above), watch the simulation and choose the options that provide optimal conditions of the simulation. These conditions include the choice of the reference frames, the time marks, the scales selected with the help of the menu items 'Zoom in' and 'Zoom out', etc. (These conditions are reproduced each time we recall this example.) Then we open the panel 'Examples' and choose 'Create new example' from the menu 'Edit'. The program suggests that we enter a title for the example and give it a brief description. Clicking the 'Ok' button, we add the new example to the end of the list. The menu item 'Move example up' allows us to put the example into a proper place in the list.

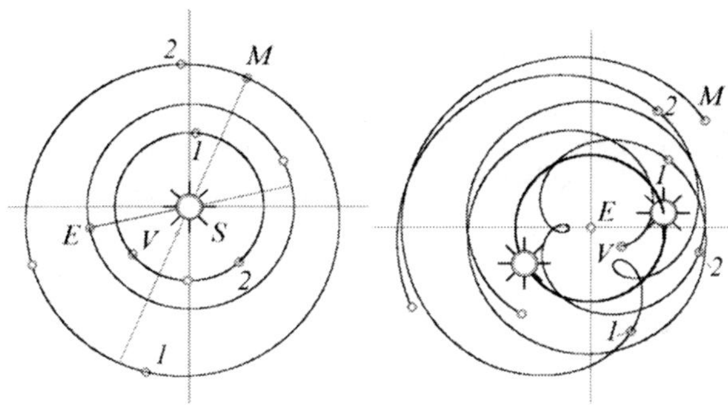


**Figure 9.1.** Three planets of the terrestrial group in the heliocentric (left) and the geocentric (right) reference frames.

To save a modified or newly created set, we choose 'Save' in the menu 'File'. The program prompts us to give a name to the set of examples (this name appears over the list of examples when we open the set) and a name (and path) to the file in which the set is to be stored. We can create as many sets as we need. To open a set afterward, we choose 'Open examples' in the menu 'File', and find the desired set by the name of the file in which the set is stored.

## 9.2 A model of the Solar System

The dimensions of the orbits of Mercury and Neptune (and Pluto) differ greatly and it is difficult to display these orbits in the same scale. However, we can separately simulate the Jovian and terrestrial planets. Figure 9.1 shows three planets of the terrestrial group—Venus $V$, Earth $E$ and Mars $M$, orbiting the Sun $S$. (For the sake of simplicity Mercury is not shown, otherwise the figure looks overcrowded.) At the initial moment all the planets are at the perihelia of their orbits. Numbers *1* and *2* mark the positions of the planets in their orbits one and two years later. At each of these moments the Earth, of course, is again at its initial position $E$. The circles without numbers show the positions of the planets at the final moment of the simulation (2.5 years).

The right panel of figure 9.1 shows the trajectories of the planets (and the Sun) in the geocentric frame. In this frame, the Sun moves in a closed elliptical (almost circular) orbit. For an observer on the Earth, during one year the Sun completes its way along the ecliptic moving counterclockwise through the zodiacal constellations. The complicated looping trajectories generated by the planets in the geocentric frame are explained by the superposition of this motion of the Sun and rather simple planetary revolutions around the Sun. When a planet passes through the apex of a small loop turned toward the Earth, for the observer on the Earth the visible motion of the planet through the stars is retrograde.

### 9.2.1 The kinematics of planetary motion

To better understand the complicated motions of the planets as seen by an observer on the Earth, next we consider the kinematics of superior and inferior planets

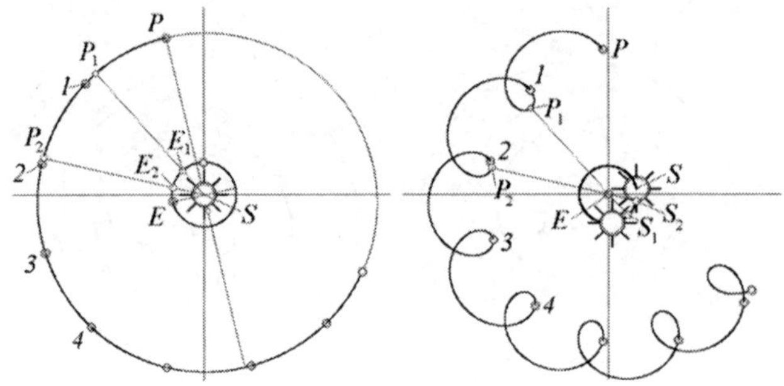

**Figure 9.2.** The Sun, Earth and Jupiter in the heliocentric (left) and the geocentric (right) frames.

separately. Figure 9.2 shows Jupiter, whose orbit is 5.2 times greater than that of the Earth. Jupiter completes one revolution along its orbit with reference to the background of the stars in 11.86 years. (This time is called the *sidereal period* of the planet.) In other words, during one revolution of Jupiter around the Sun, the Earth makes 11.86 revolutions. At the initial moment of the simulation, Jupiter and the Earth are at the perihelia of their orbits (points $P$ and $E$ respectively). The position of Jupiter in its orbit after one (terrestrial) year, two years, etc, is marked by numbers *1, 2, ….* At each of these moments the Earth returns to its initial position $E$.

When Jupiter passes through point $P_1$, the Earth is at point $E_1$ on the line joining Jupiter with the Sun. Such a collinear configuration of the superior planet with the Earth and the Sun (the planet and the Sun on the opposite sides of the Earth) is called *opposition*. In opposition the distance between the Earth and Jupiter is at its smallest. Because the Earth revolves around the Sun faster than does Jupiter, for the observer on the Earth the motion of Jupiter through the stars near opposition is retrograde.

One year after the opposition $P_1$–$E_1$, the Earth again returns to point $E_1$. But in this same time Jupiter has moved from $P_1$ forward along its orbit. Therefore the next opposition occurs after an interval somewhat longer than a year, when the Earth again passes between the Sun and Jupiter (positions $P_2$ and $E_2$). The mean time $S$ taken by the Earth and a planet to move from one opposition to the next is called the *synodic period.* This interval between successive oppositions is determined by the difference in the angular velocities of the Earth and the planet. Therefore for a superior planet $1/S = 1/E - 1/T$, where $T$ is its sidereal period and $E$ is the sidereal period of the Earth (the sidereal year, 365.257 days)[1]. For Jupiter, the sidereal period $S = 398.88$ days.

[1] The *tropical year* (the interval between successive arrivals of the Sun at the vernal equinox) is about 20 min less than the sidereal year because of the slow westward motion of the equinoxes about the ecliptic as a result of the Earth's precessional motion. The gravitational forces of the Sun and the Moon exerted on the equatorial bulge of the Earth produce a torque on the Earth. If the Earth were not spinning about its axis, it would turn under the torque so that the equatorial bulge moves toward the plane of the ecliptic. However, this torque does not change the inclination of the axis of the spinning Earth to the ecliptic, but instead under the torque the axis undergoes precession with a period of 25 800 years.

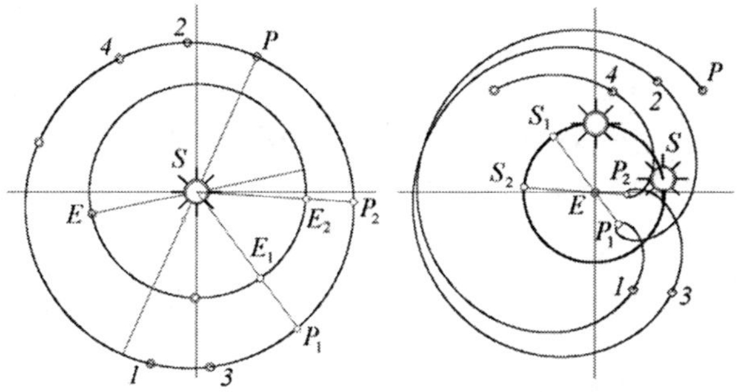


**Figure 9.3.** The Sun, Earth and Mars in the heliocentric (left) and the geocentric (right) frames.

In the geocentric reference frame (right panel of figure 9.2), the loops of Jupiter's trajectory reflect the heliocentric orbital motion of the Earth rather than the motion of Jupiter itself. The trajectory makes a loop facing the Earth each time the two planets approach one another in their heliocentric motion and align radially on the same side of the Sun. At the top of each loop Jupiter is on the side of the Earth opposite the Sun (hence the name, opposition).

The arc of Jupiter's geocentric trajectory midway between oppositions produces Jupiter's direct (counterclockwise) motion for an observer on the Earth. In the middle of the arc, Jupiter lies behind the Sun. This collinear configuration of two planets on opposite sides of the Sun is called *conjunction.* At conjunction the direct motion of Jupiter through the stars appears to us to be greatest. The angular speed of this apparent motion is generated more by the Earth's motion rather than by the motion of Jupiter itself, though Jupiter's proper motion contributes to this angular speed. The apparent motion of the Sun (caused by the Earth's revolution) is even faster because the Sun is much closer to the Earth than is Jupiter. Hence at conjunction the Sun appears to overtake Jupiter.

Of the planets whose orbits lie outside that of the Earth, Mars is nearest, its orbit being only one and a half times greater than that of the Earth (figure 9.3). In this diagram, the Earth and Mars are initially at the perihelia of their orbits ($E$ and $P$, respectively). The small circles with numbers *1, 2, …* show the position of Mars after a year, two years, etc. The first opposition occurs after more than a year, when Mars is at point $P_1$ and the Earth is at $E_1$. The difference between the angular velocities of the Earth and Mars is not as great as it is for Jupiter, so that the Earth's successive oppositions with Mars occur after an interval exceeding two years—the synodic period of Mars is 780 days. (Compare with 399 days of the synodic period of slowly moving Jupiter whose successive oppositions with the Earth repeat after an interval of only about a month longer than the terrestrial year.)

At the second opposition ($P_2$–$E_2$) the distance between the planets $E_2P_2$ is smaller than the distance $E_1P_1$ because of the comparatively large eccentricity of the orbit of Mars. For an observer on the Earth, the most favorable conditions to observe Mars occur at *great oppositions*, when Mars passes through the perihelion of its orbit, where the distance between the orbits is smallest. Great oppositions occur once every 15–17 years, approximately in August, because the Earth passes through

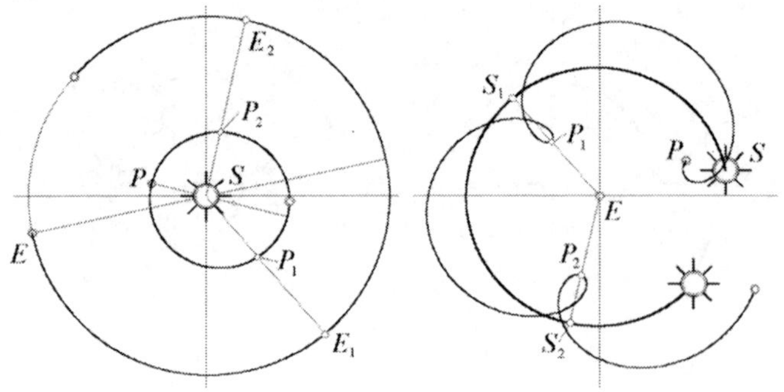


**Figure 9.4.** The kinematics of an inferior planet (Mercury) in the heliocentric (left) and the geocentric (right) frames.

the point of its orbit closest to the orbit of Mars in August. The right panel of figure 9.3 shows (in a somewhat smaller scale) the orbit of the Sun and the looping trajectory of Mars in the geocentric frame.

### 9.2.2 Kinematics of the inferior planets

The kinematics of the inner planets is illustrated in figure 9.4. The left panel shows the heliocentric orbits of Mercury and Earth. The simulation starts when the planets are at the perihelia $P$ and $E$ of their orbits. The positions of the planets at an *inferior conjunction* (the collinear configuration of the planets with the Sun in which the inner planet is between the Earth and the Sun) are marked by $P_1$ and $E_1$. (This is the second inferior conjunction; the first one occurred almost at once after the start of the simulation.) The next inferior conjunction ($P_2$–$E_2$) occurs after the Earth makes less than one third of its revolution. During this time Mercury makes one and a third revolutions in its orbit. (The synodic period is 116 days.)

The orbit of the Sun and the looping trajectory of Mercury in the geocentric frame are shown in the right panel of figure 9.4. For the observer on the Earth, the motion of Mercury among the stars near its inferior conjunctions (positions $P_1$ and $P_2$) is retrograde. At these conjunctions Mercury passes through the points $P_1$ and $P_2$. These lie nearest to the Earth and are at the 'bottom' of the Earth-facing loops. At inferior conjunctions $P_1$ and $P_2$, Mercury is on the line between the Earth $E$ and the Sun because at these moments the Sun passes through positions $S_1$ and $S_2$, respectively. In the interval between these inferior conjunctions, the motion of Mercury along the large convex arc of its geocentric trajectory is direct (counter-clockwise). In the approximate middle of this arc Mercury is again aligned with the Sun and Earth, this time behind the Sun relative to the Earth, in a position called its *superior conjunction*. At superior conjunctions the apparent direct motion of Mercury among the stars is greatest, being a combination of its orbital motion around the Sun and the apparent motion of the Sun around the Earth.

## 9.3 Hypothetical planetary systems and heavenly catastrophes

Many planets of the Solar System have moons. Even though some of the moons are as large as planets (e.g. our natural satellite, the Moon), it is difficult to

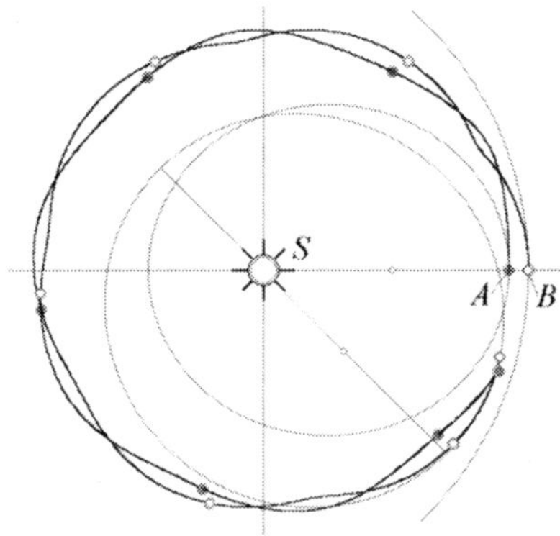


**Figure 9.5.** A binary planet orbiting a star.

simulate a real planet with a satellite because the distance planet–moon is much smaller than the planet–Sun distance. The computer screen is too small to display the orbits of a planet and its moon in the same scale. Therefore in the proposed simulations we choose exaggerated distances between a planet and its moon. The orbital motion of such a moon is more greatly perturbed by the Sun and other planets than are real moons.

We can even simulate a double planet orbiting the star, that is, a binary system like Earth–Moon, but with comparable (or even equal) masses of the components. Figure 9.5 shows the complicated intertwining trajectories traced by the components of such a double planet in the reference frame of the star. Thin lines show the ellipses that each of the components would follow in the absence of the other component under the gravitational pull of the star.

Our planetary system is a relatively calm place in the Universe. Serious catastrophes with disastrous collisions of heavenly bodies are found only in its early history. Numerous craters on the surface of the Moon and Mercury remind us of the early stages in the evolution of the Solar System. These traces of ancient collisions and bombardments by smaller bodies survived perfectly on some small planets (and natural satellites of large planets) because these celestial bodies have no atmosphere. Today, significant events like large meteoroids or new comets whose huge tails threaten to cover the Earth or other planets seldom occur in the Solar System.

The program 'Planetary System' allows us to construct an arbitrary collection of celestial bodies and thus to reproduce the events typical for a 'young' system of planets during its early evolution. For example, we can show that two planets in neighboring or intersecting orbits cannot exist for a prolonged period. Sooner or later they are found dangerously close and either collide or perturb their orbits so greatly that one of them may be ejected from the system.

When two celestial bodies of the simulated system come in contact, the program handles the event as a completely inelastic collision. That is, the two bodies join in a single body whose mass equals the sum of their masses and whose velocity immediately after the collision is determined by the conservation of momentum.

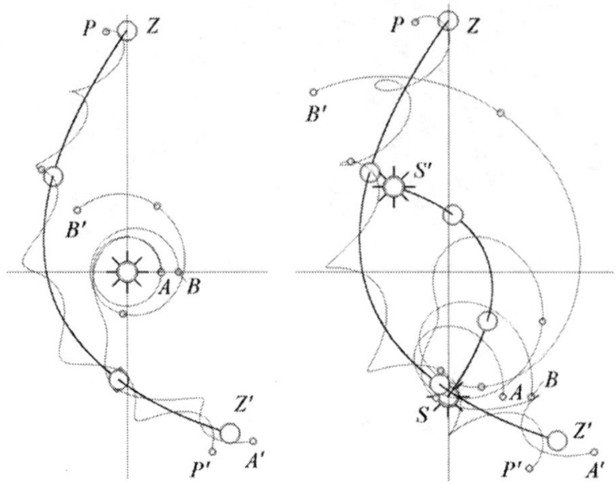


**Figure 9.6.** An encounter of two planetary systems.

When constructing planetary systems of our own, we can give free rein to our imagination. Then we can test the invented system in the simulation to discover whether its long-term behavior is stable or the system evolves through mutual collisions of the components and other catastrophes to something quite different from its initial image. In particular, we can observe and study experimentally the accumulation of larger bodies from smaller bodies via collisions in the process of the formation of planets. Various hypothetical systems that demonstrate somewhat unexpected behavior can be found in the menu item 'Examples'.

The stars in the sky look fixed. However, careful measurements show that the relative positions of these 'stationary' stars slowly change. These variations prove that the stars move in directions perpendicular to the line of vision. It is difficult to notice this motion because of the immense distances to the stars. If the distance to the star is known, we can calculate the tangential velocity of the star. On the other hand, the motion of a star along the line of vision is revealed through a uniform shift in the wavelength of its radiation (the Doppler effect). These observable radial and tangential motions of stars are explained partly by their individual movements and partly by the motion of the Sun with respect to the surrounding stars.

The Sun is found in the outskirts of the Galaxy, where stars are rare. We do not expect the Sun to encounter another star in the foreseeable future. But closer to the center of the Galaxy the concentration of stars is greater, and such events as the binary approach of stars are likely. Mutual gravitation accelerates the stars and turns them from their rectilinear trajectories. Relative to their common center of mass, the stars trace open hyperbolic Keplerian orbits. After the rendezvous the stars recede along the other asymptote of the hyperbola (figure 9.6).

We can reproduce such a stellar rendezvous with the simulation program 'Planetary System'. This interaction is especially interesting if the stars have planetary systems. The gravitational perturbations caused by a bypassing star can produce disastrous changes in the planetary system. Figure 9.6 shows a possible

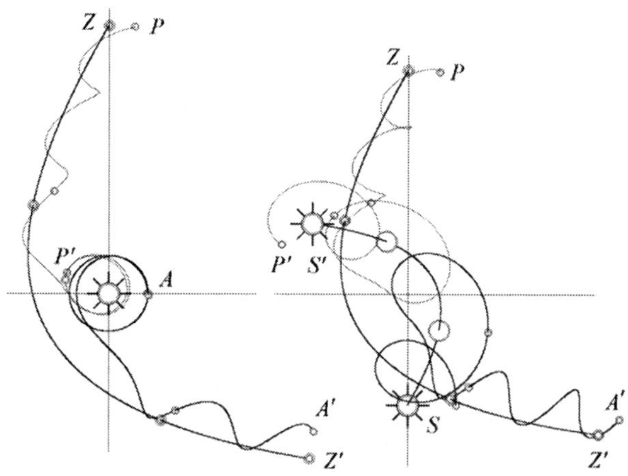


**Figure 9.7.** The encountering stars exchange planets.

scenario of the encounter of two planetary systems whose stars *S* and *Z* have masses that differ by a factor of 2. Two planets *A* and *B* revolve counterclockwise about the star *S* in almost circular orbits (initially they are found on one line with *S*) and one planet *P* revolves (clockwise) about the star *Z*.

The influence of the planets on the motion of the stars is negligible since the masses of the planets are small compared to the masses of the stars. Hence the stars move in almost hyperbolic trajectories. Portions of these hyperbolas (with a common focus at the origin) are shown on the right panel of figure 9.6.

The left panel of figure 9.6 corresponds to the frame of reference associated with the star *S*. As the 'intruder star' *Z* approaches *S*, its gravitational pull strongly perturbs the planets that orbit the star *S*. Initially almost circular, the orbit of the superior planet *B* is transformed into a large elongated ellipse. The planetary configuration during the approach of the stars is such that the influence of the intruder *Z* on the inferior planet *A* is even greater. The star *Z* captures the planet *A* and when the stars move away from one another, two planets, *P* and *A*, are orbiting the star *Z*.

The right panel of figure 9.6 shows the trajectories of all the bodies in the inertial center-of-mass frame where the stars move in homothetic hyperbolas. The circles mark the simultaneous positions of the bodies after equal time intervals. The primed letters show the positions of the bodies at the end of the simulation.

Another scenario of an encounter of two planetary systems is shown in figure 9.7. For simplicity, we consider that the star *S* is orbited by a single planet *A* and the intruder star *Z* also has a single planet *P*. (Otherwise, for stars with several planets, it is hard to follow the interlacing traces of numerous planets.) During their rendezvous the stars exchange the planets. When the stars recede, the planet *A* orbits its new host star *Z*, while the planet *P* is captured by the star *S* into a closed elliptical orbit.

The fate of encountering planetary systems is very sensitive to variations in the initial conditions. Figure 9.8 shows the same planetary systems as in the previous example, but this time the planet *P* orbits the star *Z* in a slightly different orbit. (The orbit of

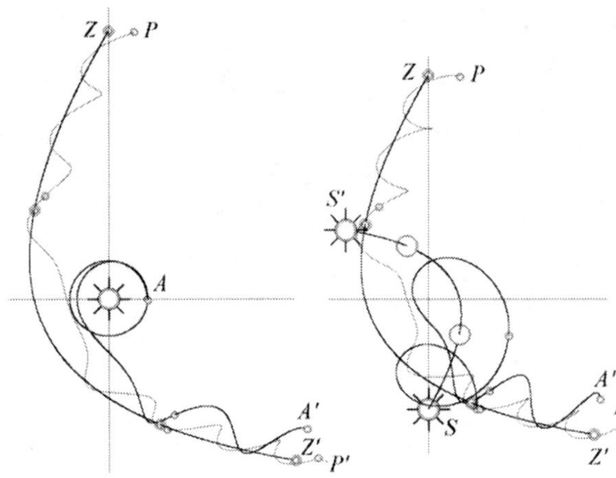


**Figure 9.8.** The 'intruder star' captures a planet.

planet $A$ around the star $S$ is unchanged.) We see that this time the intruder $Z$ again captures the planet $A$. However, the planet $P$ retains its primary and when the stars recede both planets $A$ and $P$ are orbiting the star $Z$.

## 9.4 Multiple stars

A multiple star is a group of more than two stars bound by mutual gravitation so that they move in orbits about each other. Multiple star systems of three or four stars in our Galaxy appear to be about as common as binary star systems, which are pairs of stars that are gravitationally bound to move around each other in Keplerian elliptical orbits (see chapter 7 and also section 11.9). Astronomers estimate that about half of all stars in the sky belong to either a binary or multiple star system.

The most common multiple star systems are triple stars, which typically consist of a binary pair that orbit each other at close range and a third star that orbits the binary pair at a relatively large distance. The close binary pair acts gravitationally much like a single mass on the distant companion, which is too far away to exert enough gravitational force to disrupt the stable motions of the inner pair. This arrangement is called hierarchical. If the inner and outer orbits were comparable in size, the system may become dynamically unstable, leading to a star being ejected from the system. The proportion of binary pairs that are orbited by a distant companion may be as high as 25%. The brightest star Acrux (Alpha Crucis) in the constellation Crux (or the Southern Cross) is an example of a triple star system.

The program 'Planetary System' allows us to simulate the motions of multiple star components. To do this, we consider the 'sun' to be one of the components (namely, the heaviest component) of the multiple star, and the 'planets' (whose masses can be chosen as large as the mass of the sun) to be the other components of the multiple star.

Figure 9.9 shows possible trajectories of stars in the triple star system traced in the inertial center-of-mass reference frame. Initially, all three stars lie on the same

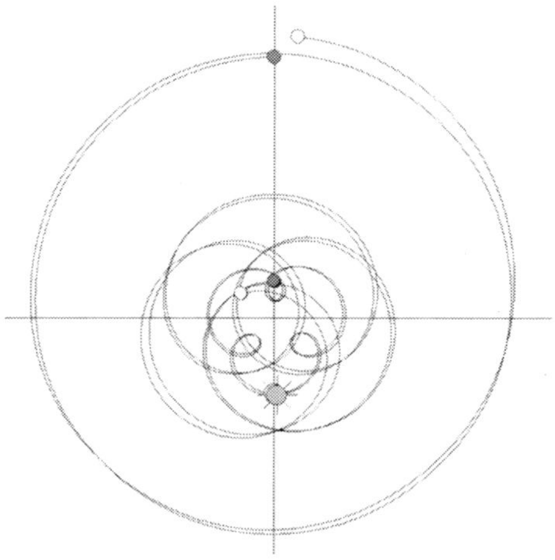

**Figure 9.9.** The trajectories of triple star components whose masses relate as 1:0.3:0.25.

straight line (vertical in the figure). The third (rather distant) star makes one revolution in an almost circular orbit around the central close pair during approximately four mutual revolutions of the binary pair. To make this simulation more impressive, the third component is chosen to be not far enough from the central pair as in real, typical triple systems. As a result, the central binary is strongly perturbed by the gravitational field of the orbiting third star. Its path, in turn, also differs considerably from a Keplerian ellipse. The simulation shows that the long-term behavior of the system is unstable.

Quadruple systems, consisting of two close pairs orbiting each other at a relatively large distance, also appear to be common, although less common than triple star systems. The Epsilon Lyrae system is a good example of a quadruple star system. The stars of Epsilon Lyrae, and most if not all other multiple star systems, move within one plane. Astronomers speculate that the only stable configurations for multiple star systems are those, like Epsilon Lyrae, in which the stars all move within one plane, but this has not been proven conclusively. Figure 9.10 illustrates the motions of quadruple star components with the help of the 'Planetary System' program. The outer pair of mutually revolving stars moves as a whole under the joint gravitational pull of the inner bound pair in a relatively large orbit.

Systems of more than four stars appear to be much less common than systems of four stars or less. One of the most famous and most complex multiple star systems is Castor, the brightest star in the constellation Gemini. In the Castor system, six stars orbit a common center of mass; two binary pairs form a quadruple system and a third pair of cool red dwarf stars orbits the inner quadruple system at a much greater distance. Complex systems such as Castor comprise only 0.1 of 1 per cent of known multiple stars.

Most real multiple stars are organized in hierarchical systems: the stars in the system can be divided into smaller groups, each of which moves in a larger orbit around the system's center of mass. Each level of the hierarchy can be treated as a two-body system by considering close pairs as if they were a single star. In these

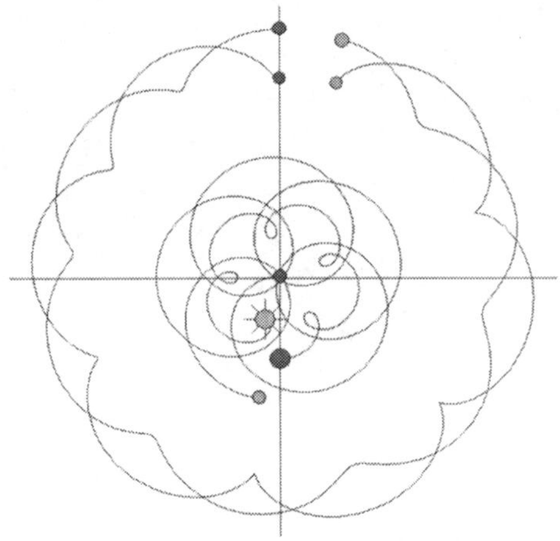

**Figure 9.10.** A quadruple star consisting of two close pairs orbiting each other.

systems the interaction between pairs is almost the same as between single stars and their motion continues along nearly Keplerian orbits around the system's center of mass.

## 9.5 Exact particular solutions to the many-body problem

For three gravitationally interacting bodies, mathematicians have found a small number of special cases in which the orbits of the three masses are periodic. In 1765, Leonard Euler discovered an example in which three masses start in a line and rotate so that they stay lined-up (see also section 8.4.3). Such a set of simple motions is unstable, however, and it would be unlikely to occur in nature. Then, in 1772, Joseph Lagrange identified a stable periodic orbit in which three masses, one of which is negligible, are located at the corners of an equilateral triangle (Lagrangian points, see section 8.4.2). Each mass moves in an ellipse in such a way that the triangle formed by the three masses always remains equilateral. A Trojan group of asteroids, which forms a triangle with Jupiter and the Sun, moves according to such a scheme.

The program 'Planetary System' allows us to simulate several curious examples of exact particular solutions to the three-body and many-body problems. Although these exact solutions are of no practical importance, their existence is interesting in principle and deserves discussion.

### 9.5.1 A star with two planets of equal mass

Figure 9.11 shows the possible simple motions in a system of a star with two planets of equal (arbitrarily large) masses. Initially the planets $A$ and $B$ are on the same straight line with the star $S$, at equal distances on opposite sides of the star. The planets have equal and opposite initial velocities in the heliocentric frame of reference (which is also the center-of-mass frame), as shown in the left panel of the figure). We see that in this symmetric configuration the motion of the system is

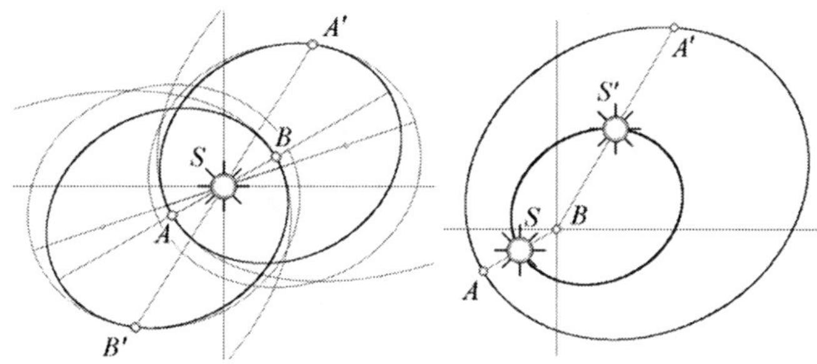


**Figure 9.11.** The simple periodic motions described by an exact particular solution to the three-body problem for a symmetric configuration of two identical planets.

regular and very simple. The star is stationary, while the planets trace closed orbits that are congruent ellipses with a common focus at the center of the star. At any moment the planets are at the opposite ends of the straight line passing through the center of the star and their velocities are equal and opposite. The planets simultaneously pass through the perihelia of their orbits, where their velocities are greatest. They also pass simultaneously through the aphelia where their velocities are smallest. After a revolution the mechanical state of the system is exactly reproduced, so the motion is periodic.

The unperturbed heliocentric elliptical orbits that each of the planets would trace in the absence of the other planet under the gravitational pull of the star are shown by thin lines in the left panel of figure 9.11. These osculating orbits that graze the actual elliptical orbits of the planets (thick lines) are shown for the perihelia $A$ and $B$ (only portions of the ellipses) and for the points $A'$ and $B'$ that are closer to the aphelia (whole ellipses). The right panel of figure 9.11 shows the trajectories of the sun $S$ and planet $A$ in the reference frame of planet $B$ (in a somewhat smaller scale).

This simple behavior of the three-body system can easily be explained. In the symmetric configuration, the gravitational forces exerted on the star by the planets are equal and opposite, so that the star is stationary in the equilibrium position until the symmetric configuration is violated. The gravitational forces exerted on each planet by the star and the other planet are both directed toward the center of the star because the other planet is on the same line with the star. Hence the resulting force is central. We can show (see section 11.11) that its magnitude is inversely proportional to the square of the distance between the star and the planet. Therefore we can consider the planet to move in a *stationary* Newtonian inverse-square gravitational field whose source is located at the center of the star. The effective mass of the stationary source $M_{\text{eff}}$ is somewhat greater than that of the star by virtue of the additional gravitational pull of the other planet ($M_{\text{eff}} = M + m/4$, where $M$ is the mass of the star and $m$ is the mass of either of the planets). In this effective gravitational field the planet traces a closed Keplerian ellipse. The second planet moves in an equivalent effective gravitational field and traces a congruent ellipse.

The simple situation described above holds only for the symmetric configuration of the system. This configuration is preserved during the motion provided the initial

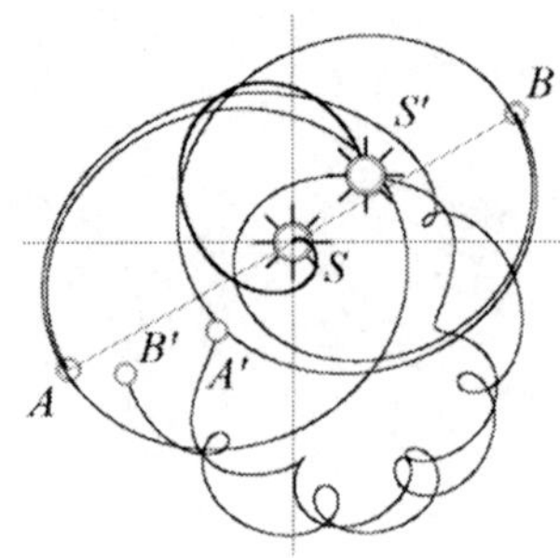


**Figure 9.12.** Instability of the periodic motion described by the exact particular solution of the three-body problem.

velocities of the planets relative to the star are exactly equal and opposite. If the velocities slightly differ in magnitude or direction, or the distances from the star to the planets are not exactly equal, or the three bodies do not lie exactly on the same straight line, the paths of the planets sooner or later deviate from Keplerian ellipses and these deviations progressively increase. The motion of the system becomes irregular (chaotic) and very complicated.

Figure 9.12 illustrates a possible evolution of events in the case of slightly unequal initial distances of planets $A$ and $B$ from the star $S$. These complicated long-term trajectories change drastically if we enter a tiny variation in the initial conditions. Hence the periodic motion described by the exact particular solution of the three-body problem is unstable.

### 9.5.2 A 'round dance' of identical planets

Similar periodic exact solutions in which the bodies trace closed Keplerian orbits exist for systems of several bodies of equal masses surrounding a central body. Let $n$ bodies of equal masses ('planets') be located at all $n$ vertices of a regular (equilateral) polygon and one more body (a 'star' whose mass can differ from the masses of the other bodies) be located at the center of the polygon.

In this symmetric configuration the central body is in equilibrium under the joint gravitational pull of all other bodies. The resulting gravitational force exerted on any of the other bodies (on a 'planet') by the central body and by the other planets is directed toward the center and its magnitude is inversely proportional to the square of the distance from the center (or, which is the same, to the square of a linear dimension of the polygon, e.g. of the length of its side).

Therefore the 'planets' can trace congruent Keplerian ellipses with the common focus at the 'star', provided the initial velocities of the planets are equal in magnitude and make equal angles with the corresponding radius vectors of the planets. The symmetric polygonal configuration of the system is preserved during the motion (figure 9.13).

In particular, the 'planets' can move uniformly at equal distances from one another along the same circular orbit (circumscribed about the polygon). In this case the polygon, with the planets at its vertices, rotates uniformly about its center.

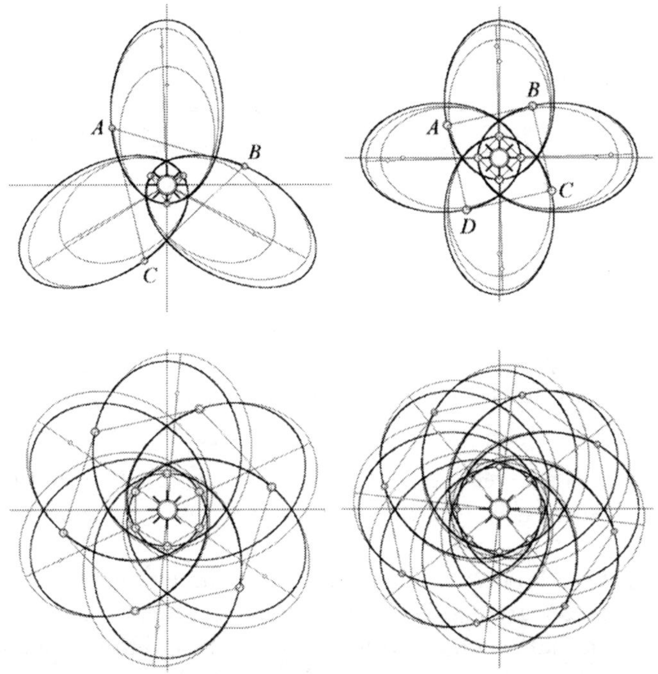


**Figure 9.13.** The polygonal systems of identical massive bodies surrounding the central body in symmetric motions described by exact particular solutions to the many-body problem.

For elliptical trajectories of the planets, the angular velocity of the polygon is greatest when the planets pass simultaneously through the perihelia of their orbits. In this non-uniform rotation of the polygon, the lengths of its sides vary periodically.

The upper panels of figure 9.13 show examples of these exact solutions for systems of three (left) and four 'planets' (right). Moving along elliptical trajectories, at any moment the bodies are at the vertices of a regular triangle and a square, respectively. Thin lines show the unperturbed orbits that the 'planets' would trace in the absence of the other 'planets' under the gravitational pull of the 'star' (about the center of mass of the two-body system consisting of the star and the single planet). These osculating orbits are shown for the perihelia of the actual orbits and for the moments at which the 'planets' pass through the points marked by small circles.

The lower panels of figure 9.13 show similar systems of six and eight 'planets' of equal mass orbiting the 'star' in symmetric equilateral configurations. The regular polygon (at whose vertices the 'planets' are found) rotates non-uniformly and the lengths of its sides vary periodically during the rotation. The osculating ellipses shown by thin lines correspond here to the unperturbed orbits of individual 'planets' in the frame of the 'star' (rather than in the center-of-mass frame).

### 9.5.3 Keplerian motions in equilateral configurations

We note that in the exact solutions to the many-body problem considered above, the mass of the central body can be zero. That is, a system of $n$ bodies of equal masses

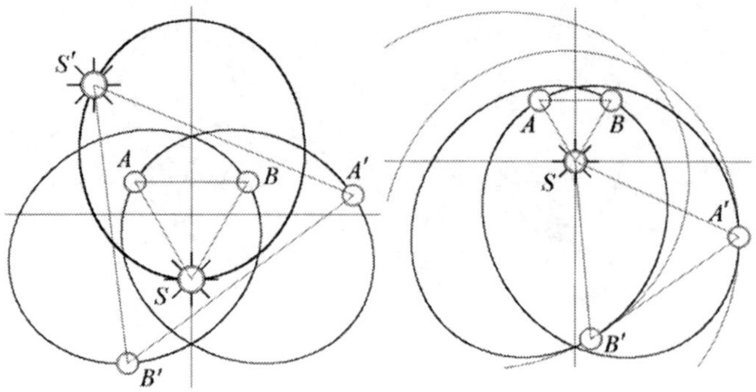


**Figure 9.14.** The regular motions of three bodies of equal masses in the equilateral configuration. Trajectories are shown in the inertial center of mass frame of reference (left) and in the frame of reference associated with one of the bodies (right).

located at the vertices of a regular $n$-sided polygon, under their mutual gravitational attraction, can perform a beautiful 'round dance' even in the absence of a central body.

In particular, three bodies of equal masses in the equilateral configuration can synchronously trace congruent ellipses whose major axes make angles of 120° with one another. To simulate such a motion with the program 'Planetary System', we should choose two 'planets' that form with the 'star' an equilateral initial configuration and enter the masses of the 'planets' equal to the mass of the star. The heliocentric initial velocities of the 'planets' should be equal in magnitude and form equal angles with the radius vectors of the 'planets'.

Figure 9.14 shows the orbits of the three bodies $A$, $B$ and $S$ of equal masses in the center-of-mass reference frame (left panel) and in the 'heliocentric' reference frame associated with $S$ (right panel, where the scale is somewhat smaller). The thin lines grazing the actual trajectories show portions of the heliocentric orbits that each of the bodies $A$ and $B$ would have traced in the absence of the other (that is, only under the gravitational pull of the 'star' $S$) for the moment at which the planets pass through points $A'$ and $B'$.

In the symmetric polygonal configuration with arbitrary number $n$ of the vertices, the resulting gravitational force exerted on any of the bodies by the other bodies of equal masses that are located at the vertices of the polygon is directed toward its center. This immediately follows from the symmetry of the system. The magnitude of this net force is inversely proportional to the square of the distance of any body from the center of the polygon (or, which is the same, to the square of a linear dimension of the polygon, e.g. of the length of its side).

Therefore the 'planets' can trace congruent Keplerian ellipses with the common focus at the center of the polygon, provided the initial velocities of the planets are equal in magnitude and make equal angles with the corresponding radius vectors of the planets. The symmetric polygonal configuration of the system is preserved during the motion.

In particular, the 'planets' can move uniformly at equal distances from one another along the same circular orbit (circumscribed about the polygon). In this case the

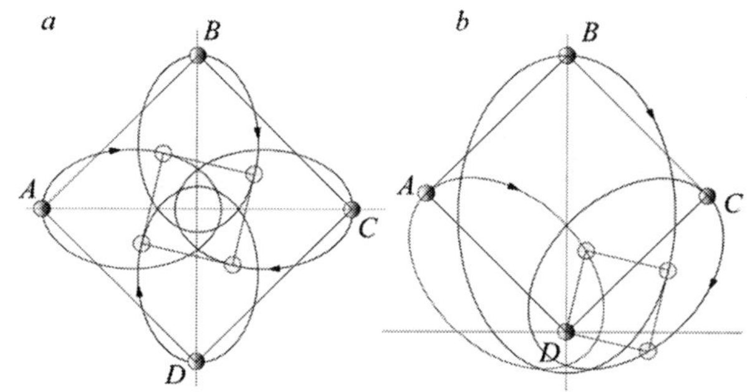


**Figure 9.15.** The regular motions of four bodies of equal masses in the equilateral (square) configuration. Trajectories are shown in the inertial center-of-mass frame of reference (*a*) and in the frame of reference associated with one of the bodies (*b*).

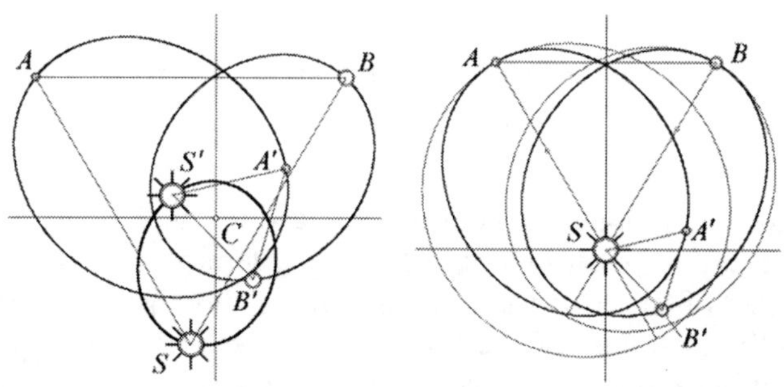


**Figure 9.16.** The regular motions of three bodies of unequal masses in the equilateral configuration. Trajectories are shown in the inertial center-of-mass frame of reference (left) and in the frame of reference associated with one of the bodies (right).

polygon, with the planets at its vertices, rotates uniformly about its center. For elliptical trajectories of the planets, the polygon rotates non-uniformly: the angular velocity of the polygon is greatest when the planets pass simultaneously through the perihelia of their orbits and smallest at the aphelia. In this non-uniform rotation of the polygon, the lengths of its sides vary periodically. An example of such regular motion of four bodies of equal masses located at the vertices of a square is shown in figure 9.15.

The equilateral configuration of three bodies is especially interesting because it can be preserved during the motion even when the masses of the bodies are different (figure 9.16). It can be shown (see section 11.11.2) that the total gravitational force exerted on each of the bodies by the other two bodies is directed toward the center of mass of the system and is inversely proportional to the square of the distance from the center of mass. It can also be shown that the accelerations of the bodies produced by these forces are in the same ratio as are the distances of the bodies from the center of mass. Therefore, the initial equilateral configuration can be preserved during the motion, provided the initial velocities are chosen properly.

In other words, in the equilateral configuration of three bodies coupled by the gravitational forces, each of the bodies can be considered as moving in an effective *stationary* central inverse-square gravitational field with the source at the center of

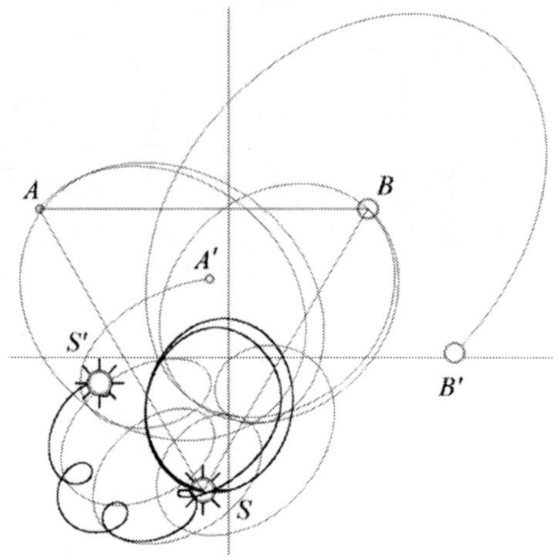


**Figure 9.17.** The transition to irregular motion in a system whose initial motion is close to that described by the exact solution.

mass of the system, although this field is produced by the *moving* bodies. Hence the bodies can trace synchronously homothetic Keplerian ellipses with the common focus at the center of mass of the system. Linear dimensions of these ellipses are proportional to the distances of the bodies from the center of mass.

To simulate this motion, we choose an equilateral initial configuration of the bodies and enter certain initial velocities. An example of such a simple periodic motion is shown in figure 9.16 ($m_A = 0.3m_S$, $m_B = 0.6m_S$). In the inertial center-of-mass frame (left) the bodies trace homothetic elliptical orbits of different sizes and orientations.

In the 'heliocentric' frame associated with the body $S$ of greatest mass (right panel of figure 9.16), the bodies $A$ and $B$ trace the congruent ellipses shown by thick lines. The major axes of these ellipses form an angle of 60°. The thin lines show the (non-congruent) heliocentric osculating orbits that each of the bodies $A$ and $B$ would have traced around $S$ in the absence of the other body (for the moment at which $A$ and $B$ pass through the aphelia of their orbits). This regular periodic motion of the three bodies is unstable with respect to (small) variations in the initial conditions that disturb the symmetry of the system. This instability of motion in the initially equilateral configuration of the bodies is illustrated by figure 9.17.

Nature seems to deal with the three-body problem in a simple manner for many stellar systems in our Galaxy consisting of three or more stars bound by their mutual gravitational attraction. In all of these systems, the objects seem to degenerate to a hierarchical succession of two-body problems. For example, should the system contain three stars, two will be tightly bound, orbiting as one would expect from the two-body solution while the third will be found at a distance corresponding to many times the separation of the close orbiting pair. Four gravitationally bound stars always appear as a binary of binaries, and so forth. It is generally believed that there are no stable orbits involving three comparable masses with comparable separations.

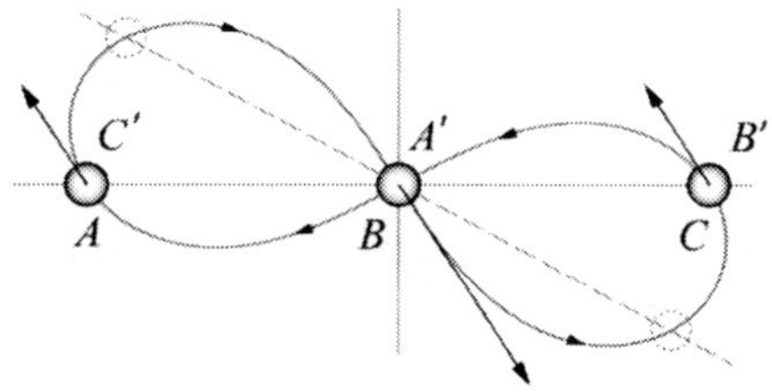


**Figure 9.18.** Regular periodic motion of three bodies of equal masses along a figure-eight highly symmetric orbit.

### 9.5.4 A remarkable three-body motion along a figure-eight

At the very beginning of the 21st century mathematicians Richard Montgomery and Alain Chenciner added another unexpected beautiful periodic solution to the equations of motion for three gravitationally interacting bodies. They discovered a surprisingly simple periodic orbit for the Newtonian three-body problem: three bodies of equal masses chase each other around a fixed planar highly symmetric closed orbit in the shape of a figure-eight. From many points of view this is the simplest periodic solution for the problem, after those of Lagrange and Euler described in section 9.5.3 and section 11.10 (see also section 11.11).

The motion is characterized by zero angular momentum and a very rich symmetry pattern. The bodies ($A$, $B$, $C$ in figure 9.18) start from an equally spaced collinear configuration. The velocities of bodies $A$ and $C$ are equal in magnitude and have the same direction; the velocity of the middle body $B$ is twice as great and oppositely directed. The bodies exactly reproduce this configuration after a certain time interval $T$, during which each body traces one and the same closed figure-eight curve. After a $T/3$ time interval (one third of the total period), the bodies again occur on the same straight line as in the initial configuration, but in a different order ($C'$, $A'$, $B'$ in figure 9.18). There is one more straight line (dashed line in figure 9.18) on which the bodies simultaneously occur at a $T/6$ time interval after the first collinear configuration. This dashed line makes the same angle as the straight line of the initial collinear configuration $A$, $B$ and $C$ with respect to the figure-eight symmetry axis (to the opposite direction). In the meantime, between these two collinear configurations (after $T/12$) the bodies occur in an isosceles triangular configuration.

The most surprising feature of the figure-eight motion is its dynamical stability: the bodies will stay on track if perturbed a little. Indeed, this type of motion survives a tiny disturbance without serious disruption. Figure 9.19 illustrates this stability: the motion loses its periodicity—each next-in-turn pattern slightly differs from the preceding one, but the general character of the motion is preserved.

Moreover, the figure-eight motion, being exactly periodic for the bodies of equal masses, persists in a similar fashion over a range of masses, when the three masses are not precisely the same. This means there is a chance that the figure-eight orbit might actually be seen in some stellar system. However, it is a pretty small chance—somewhere between one per galaxy or even one per universe!

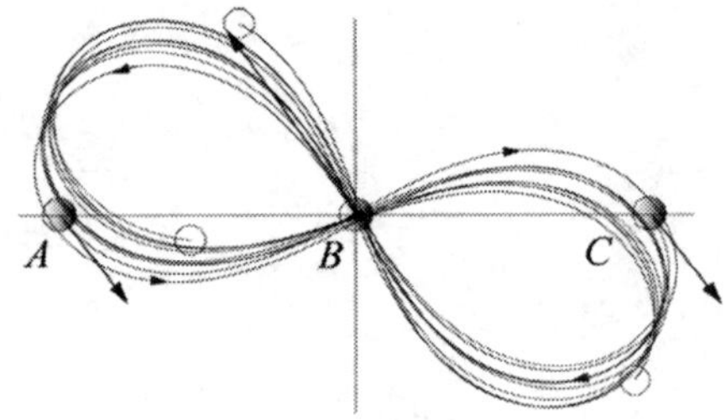


**Figure 9.19.** Numerically simulated almost periodic motion of three bodies of equal mass after a slight disturbance of the regular figure-eight motion.

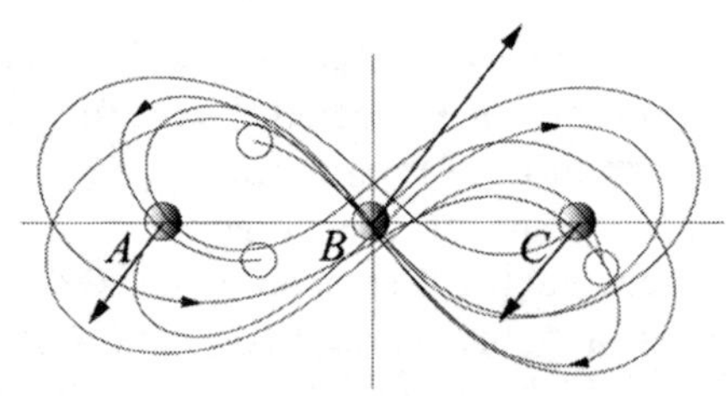


**Figure 9.20.** Chaotic figure-eight motion of three bodies of equal mass after a strong disturbance.

The general character of the figure-eight motion is preserved even after rather strong disturbances: computer simulations show that the bodies continue chasing each other indefinitely in the same manner within a restricted spatial domain without collisions, as shown in figure 9.20.

The existence of the three-body, figure-eight orbit has prompted mathematicians to look for similar orbits involving four or more masses. For the time being, they have found hundreds of exact solutions for the case of $N$ equal masses traveling a fixed planar curve. However, all such motions are not stable and thus are of no practical significance.

## Questions and problems

1. (*) **Circular orbits of massive planets.** Two planets of equal masses $m$ are positioned on the opposite sides of the star of mass $M$ at equal distances from it. What initial velocities must the planets have in order to move around the star in a circular orbit, remaining at all times at the opposite ends of its diameter? Express the velocity in units of the unperturbed circular velocity (i.e. the circular velocity of one of the planets in the absence of the other). Verify your answer using the simulation experiment for $m = M/2$.
2. (*) **Solid rotation of three identical bodies in the equilateral configuration.** Three bodies of equal masses $m$ are positioned at the vertices of an equilateral triangle whose side equals $a$. What initial velocities must the bodies have in order to move circularly under the forces of mutual gravitational attraction? What is the angular velocity of such a solid rotation of the system? Try to reproduce this motion in the simulation experiment. Is the motion stable?

3. **Rotation of four identical bodies in the equilateral (square) configuration.** Four bodies of equal masses $m$ are positioned at the vertices of a square whose side equals $a$. What initial velocities must the bodies have in order to move circularly under the forces of mutual gravitational attraction? What is the angular velocity of such a solid rotation of the system? Try to reproduce this motion in the simulation experiment.
4. (**) **Bodies of unequal mass in the equilateral configuration.** Prove analytically that three bodies of arbitrary masses $m_1$, $m_2$ and $m_3$ can move synchronously in Keplerian elliptical or circular orbits, preserving the equilateral configuration during the motion. Express the period of this motion in terms of the masses of the bodies, the distance $r$ between them in the initial equilateral configuration and their initial velocities $v_0$.
5. **'Round dance' of planets in a symmetric planetary system.** Imagine a hypothetic system consisting of $n$ identical planets located at the vertices of an $n$-sided regular polygon, the center of which is occupied by a star. Explain why in this system the planets can move along Keplerian ellipses, just as if each of the planets were moving under the sole central force created by a virtual immovable source located at the center of the polygon, and the magnitude of this single force were inversely proportional to the distance from the source. We note that actually each planet in this system is subjected to gravitational attraction not only by the immovable star, but also by all other moving planets of this system. What initial velocities must the planets in this configuration have in order that we can observe such a round dance? Is this regular motion stable? Is such a round dance possible in the absence of the central body (the star)?

# Part II

## The simulated phenomena

**IOP** Publishing

Motions of Celestial Bodies
Computer simulations
**Eugene Butikov**

# Chapter 10

# Phenomena and concepts in celestial mechanics—an introductory approach

Celestial mechanics and its modern branch—space dynamics or astrodynamics—study the motion of various celestial bodies such as stars, planets, comets, natural and artificial satellites, and spacecraft. The theoretical background of celestial mechanics is given by Newton's laws of motion and Newton's law of gravitation.

The three laws of motion that are the foundation of Newtonian mechanics are applicable to all macroscopic bodies moving at non-relativistic velocities (small compared to the speed of light) under forces of any kind. For massive celestial bodies, the most important are the forces of gravitational attraction, described by Newton's law of gravitation. This chapter deals with important properties of Keplerian motions which can be investigated on the basis of Newton's law of gravitation and Newton's laws of motion with the use of rather modest mathematical means. The next chapter 11 is much more sophisticated and is intended for in-depth study of the subject.

## 10.1 Newton's law of universal gravitation

The central force that makes a planet orbit its sun obeys Newton's law of universal gravitation. Newton's law of gravitation states that any two material points (point masses) in the Universe attract each other with a force proportional to their masses $m_1$ and $m_2$ and inversely proportional to the square of the distance $r$ between them:

$$F = G\frac{m_1 m_2}{r^2}. \tag{10.1}$$

Here $G$ is the universal *gravitational constant*, whose value depends on the chosen units of force, length and mass. In the International System of units (SI) $G = 6.672 \cdot 10^{-11}\ \mathrm{N\,m^2\,kg^{-2}}$.

For massive bodies of finite dimensions, Newton's law of gravitation expressed by (10.1) is valid when the distance between the bodies is much greater than their linear dimensions. The bodies can then be regarded as point masses. However, it is possible to prove from (10.1) and the principle of superposition (the vector character of forces) that the gravitational interaction between bodies whose distribution of mass is spherically symmetric obeys the same equation (10.1) for any distances between the bodies. In other words, real bodies with spherical symmetry interact as point masses positioned at the geometric centers of the bodies. The distribution of mass in stars and planets is very nearly spherically symmetric. Consequently, the gravitational field created by these bodies is the same as if their masses were concentrated at their centers and (10.1) can be used to calculate the gravitational force acting on their satellites even in cases in which the satellites approach close to the surface of the primary. The distance $r$ in (10.1) in such cases is the distance between the satellite and the center of the primary.

In particular, (10.1) is applicable to any small bodies located both over the Earth and on its surface. And it does not matter whether this body, say, an artificial satellite, is itself spherically symmetric because the gravitational field of the Earth is practically the same at all points of the body. In other words, within a satellite's extent, the gravitational field of the Earth is almost uniform and we can assume that the gravitational force is applied at the center of gravity of the satellite.

To emphasize that the gravitational force is a central force and is directed toward the center of the planet, we can rewrite (10.1) for the vector $\mathbf{F}(\mathbf{r})$ of force experienced by a body of mass $m$ at a given point $\mathbf{r}$ in the gravitational field of a planet of mass $M$:

$$\mathbf{F}(\mathbf{r}) = -G\frac{Mm}{r^2}\frac{\mathbf{r}}{r}. \qquad (10.2)$$

The radius vector $\mathbf{r}$ is the displacement of the body relative to the center of the planet. Equation (10.2) shows explicitly that the gravitational force $\mathbf{F}(\mathbf{r})$ is everywhere directed opposite the unit vector $\mathbf{r}/r$, which points from the center of the planet toward the given point $\mathbf{r}$.

Since the gravitational force is radial, it is convenient to introduce the notation $F_r$ for the projection of the force onto the direction of the radius vector $\mathbf{r}$:

$$F_r(r) = -G\frac{Mm}{r^2}. \qquad (10.3)$$

The radial projection $F_r$ is negative because the force $\mathbf{F}(\mathbf{r})$ is directed toward the center of the planet.

The motion of celestial bodies is governed first of all by the Newtonian gravitational forces described by (10.1) and (10.2).

Near the surface of the Earth, where the distance $r$ in (10.3) approximately equals the Earth's radius $R \approx 6370$ km, the gravitational force makes any free body (independent of its mass and other properties) fall with an acceleration $g \approx 9.81$ m s$^{-2}$ (the *acceleration of free-fall*):

$$g = G\frac{M}{R^2}, \quad \text{whence} \quad GM = gR^2. \qquad (10.4)$$

When we deal with motions of satellites orbiting the Earth, it is convenient to replace, with the help of (10.4), the product $GM$ with $gR^2$ in (10.2) for the gravitational force. This substitution allows us to avoid memorizing the values of the Earth's mass $M$ and the gravitational constant $G$ for calculations.

The Earth exerts a force on any body located over the Earth although there is no physical connection between the Earth and the body. We say that the presence of the Earth produces a *gravitational field.* Any particle in this field experiences a force directed toward the center of the Earth. The force exerted on a particle of unit mass is called the *strength* of the gravitational field. The strength of the field decreases with distance according to the inverse-square law. The gravitational force experienced by a particle is proportional to its mass and to the strength of the field at the particle's location. Near the surface of the Earth the strength of the gravitational field is equal to the acceleration of free-fall **g**.

We should realize that Newton's law of gravitation, although it characterizes the field quantitatively, nevertheless gives no physical explanation of gravity. The field is an elegant way to describe gravity, but it does not tell us what gravity is.

## 10.2 Potential energy of a body in the Newtonian gravitational field

The gravitational field is a *potential,* or *conservative field.* That is, the work performed by a gravitational force when a body is moved from some point to another point does not depend on the path along which the body moves. The *potential energy* of a body at some point in the gravitational field is measured by the work performed by the gravitational force when the body is transferred from this point to a point at which the potential energy is assumed to be zero. This point of zero potential energy can be chosen arbitrarily. Usually either a point on the surface of the planet (of the Earth in particular) or a point at infinity is chosen to be the point of zero potential energy. This arbitrariness does not influence any physical applications of the potential energy because only the difference in its values is important.

In celestial mechanics it is convenient to choose the point of zero potential energy at infinity, where the gravitational force vanishes. With this choice, the potential energy $U(r)$ of some body of mass $m$ at a finite distance $r$ from the center of a planet of mass $M$ is negative because the work performed by the force of gravitational attraction, (10.1), is negative when the body is transferred from this point to infinity. The value of the gravitational potential energy depends only on distance $r$ from the center of the planet, that is, the field is spherically symmetric:

$$U(r) = -G\frac{Mm}{r}. \tag{10.5}$$

For the potential energy of a body in the gravitational field of the Earth, we can, with the help of (10.4), rewrite (10.5) as follows:

$$U(r) = -mg\frac{R^2}{r}. \tag{10.6}$$

The familiar expression $U(h) = mgh$ for the potential energy of a body at a height $h$ over the Earth's surface is approximate and valid only for small values of $h$ compared to the Earth's radius $R$ ($h \ll R$). In other words, the expression $U(h) = mgh$

holds in the approximation of the 'flat Earth', that is, within small enough spatial regions for which the gravitational field of the Earth can be considered as uniform.

The approximate formula $U(h) = mgh$ can be obtained from the exact equation (10.6), if we express there the distance $r$ of the body from the Earth's center as the sum $R + h$, expand $U(r)$ in a power series and keep there the terms to the first power of the small parameter $h/R$:

$$U(h) = -mg\frac{R^2}{R+h} = -mg\frac{R}{1+h/R} \approx -mgR\left(1 - \frac{h}{R}\right) = -mgR + mgh. \quad (10.7)$$

Then we should discard the constant term $-mgR$ (it has the physical sense of the potential energy of the body at the surface) because in the expression $U = mgh$ the potential energy on the surface is assumed to be zero.

## 10.3 Circular velocity and escape velocity

Circular orbits can exist in any central gravitational field. For a given distance $r$ from the center of the force, the body, in order to move in a circle, must have a definite velocity that is perpendicular to the radius vector. The value of this *circular velocity* $v_c$ can be calculated with the help of Newton's second law by equating the centripetal acceleration $v_c^2/r$ for motion along a circle of radius $r$ to the acceleration $GM/r^2$ created by the gravitational force:

$$v_c = \sqrt{\frac{GM}{r}}. \quad (10.8)$$

The circular velocity is inversely proportional to the square root of the orbit's radius and is independent of the satellite's mass. This dependence of the circular velocity on the radius of the orbit is characteristic of the inverse-square law of the gravitational force.

The period of revolution around the planet of mass $M$ along a circular orbit of radius $r$ can be found by dividing the circumference $2\pi r$ of the orbit by the constant value of the circular velocity $v_c$:

$$T = \frac{2\pi r}{v_c} = 2\pi\sqrt{\frac{r^3}{GM}}. \quad (10.9)$$

As we can see from (10.9), the square of the period of revolution is proportional to the cube of the radius of the orbit (Kepler's third law for the special case of circular orbits) and inversely proportional to the planet's mass. An analytic proof of Kepler's third law for the general case of elliptical orbits is considered in section 11.3.

The dependence of the period of revolution $T$ on the planet's mass $M$ gives a simple and precise method of 'weighing' a planet (more exactly, a method of determining the planet's mass) by measuring the periods of revolution of the planet's satellites.

The *kinetic energy* of a satellite orbiting a planet along a circular orbit can be expressed, with the help of (10.8), through the radius of the orbit:

$$E_{\mathrm{kin}} = \frac{1}{2}mv_{\mathrm{c}}^2 = \frac{GmM}{2r}. \tag{10.10}$$

Comparing this expression with (10.5), we see that the kinetic energy equals one half the magnitude of the satellite's potential energy $U(r)$.[1] Hence, the total energy $E = E_{\mathrm{kin}} + U(r)$ is negative and equals the kinetic energy in magnitude:

$$E = -E_{\mathrm{kin}} = -\frac{1}{2}mv_{\mathrm{c}}^2 = -\frac{GmM}{2r}. \tag{10.11}$$

The absolute value of the total energy of a satellite in a circular orbit is inversely proportional to the radius $r$ of the orbit.

For the Earth's satellites, we can replace the product $GM$ with $gR^2$ in (10.8) and obtain the value $v_{\mathrm{c}} = \sqrt{gR^2/r}$ for the circular velocity. For a hypothetical ground-level circular orbit $r = R$ and so $v_{\mathrm{c}} = \sqrt{gR} \approx 7.9\,\mathrm{km\,s^{-1}}$. This value is sometimes called the *first cosmic velocity*.

The *escape velocity* $v_{\mathrm{esc}}$ for a given distance $r$ from the center of a planet is the minimal speed that a body must have at this point in order to overcome the gravitational attraction of the planet and to recede to infinity. The value of the escape velocity can be found from the law of energy conservation. The minimal speed of a body at a distance $r$ needed to escape the gravitational field corresponds to a speed of zero at infinity and hence to a kinetic energy of zero at an infinite distance from the planet. The value of the potential energy at infinity is also zero. Hence the total energy of a body that is to escape is zero. With the help of (10.9) for the potential energy, we can write:

$$\frac{mv_{\mathrm{esc}}^2}{2} - \frac{GmM}{r} = 0,$$

whence

$$v_{\mathrm{esc}} = \sqrt{\frac{2GM}{r}}. \tag{10.12}$$

Comparing this expression with (10.8), we see that the escape velocity $v_{\mathrm{esc}}$ for any distance $r$ from the center of a planet is $\sqrt{2} \approx 1.41$ times greater than the circular velocity. Its value is independent of the mass $m$ of the body. For a body at the surface of the Earth, the escape velocity $v_{\mathrm{esc}} = \sqrt{2gR} \approx 11.20\,\mathrm{km\,s^{-1}}$ (the *second cosmic velocity*).

[1] The same relation between the *averaged-over-a-period* values of the kinetic and potential energies of a satellite is also valid for elliptical orbits.

If the magnitude of the initial velocity $v_0$ of a body equals the escape velocity, the body escapes the gravitational field of the planet and recedes to infinity independent of the direction of the initial velocity, provided its trajectory does not intersect the surface of the planet. The motion occurs along a parabolic trajectory or along a straight line. The latter case corresponds to an upward direction of the initial velocity.

If the initial velocity of a body exceeds the escape velocity, the body recedes to infinity along a hyperbolic trajectory. At an infinitely large distance from the planet, the motion of the body is uniform and rectilinear, occurring along one of the asymptotes of the hyperbola. The constant value of the velocity $v_\infty$ of this motion can be found from the law of energy conservation:

$$\frac{mv_0^2}{2} - \frac{GmM}{r} = \frac{mv_\infty^2}{2},$$

whence

$$v_\infty = \sqrt{v_0^2 - \frac{2GM}{r}} = \sqrt{v_0^2 - v_{\text{esc}}^2}. \qquad (10.13)$$

This value is called the remaining velocity or the *hyperbolic excess of velocity.*

## 10.4 Geometric properties of Keplerian orbits

In the general case the motion of a body under the action of a central Newtonian gravitational force whose magnitude is inversely proportional to the square of the distance, occurs along one of the *conic sections*: an ellipse (or, in particular, a circle), a parabola, or a hyperbola. It is convenient to write the equation of a trajectory of a Keplerian motion in polar coordinates:

$$r(\varphi) = \frac{p}{1 + e\cos\varphi}. \qquad (10.14)$$

Here $r(\varphi)$ (the length of the radius vector) is the distance of a point on the trajectory from the center of force (a focus of the conic section) and $\varphi$ is the angle between the radius vector and the axis of symmetry of the trajectory (the major axis of the conic section) directed from the focus toward the nearest point of the trajectory. The quantity $p$ in (10.14) has a dimension of length. It is called the *semilatus rectum* or the *orbital parameter* of the conic section. The dimensionless quantity $e$ in (10.14) is called the *eccentricity* of the conic section.

For $e = 0$ (10.14) gives $r = p$, that is, the distance $r$ does not depend on $\varphi$. Hence for $e = 0$ the orbit is a circle whose radius equals $p$. Otherwise, for $e > 0$, the orbital parameter $p$ is the distance between the center of force and the point of the orbit determined by the angle $\varphi = \pm\pi/2$, for which $\cos\varphi = 0$. Geometrically, $p$ is the length of the perpendicular to the major axis, drawn from the focus to the orbit. For $e < 1$ (10.14) describes an ellipse (figure 10.1), for $e = 1$ a parabola and for $e > 1$ a hyperbola.

An elliptical orbit can also be characterized by the distances $r_P$ and $r_A$, where $r_P$ is the distance between the center of force and the nearest point $P$ of the ellipse and

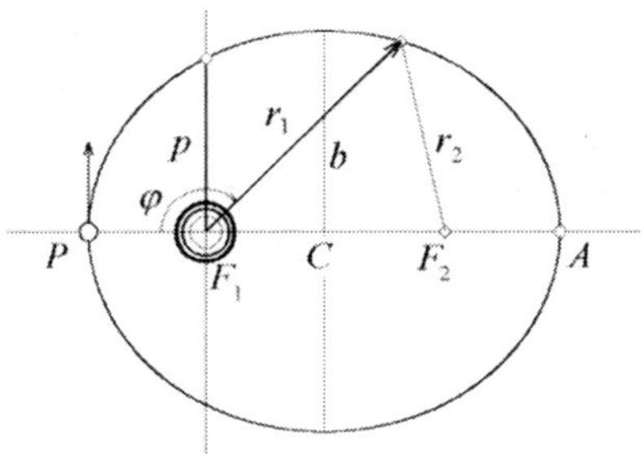


**Figure 10.1.** Elliptical trajectory of a body moving in a Newtonian central gravitational field.

$r_A$ is the distance between the center of force and the farthest point $A$ of the ellipse (see figure 10.1). Point $P$ is called the *perihelion* for a planetary orbit, the *perigee* for the orbits of Earth's satellites and the *pericenter* for the general case. Similarly, $A$ is called the *aphelion, apogee*, or *apocenter*. In (10.14) the value $\varphi = 0$ defines the point $P$, while $\varphi = \pi$ defines the point $A$. Substituting these values into (10.14), we find:

$$r_P = \frac{p}{1 + e} \qquad r_A = \frac{p}{1 - e}. \tag{10.15}$$

We can express the eccentricity $e$ of the orbit in terms of $r_P$ and $r_A$ from these relations as:

$$e = \frac{r_A - r_P}{r_A + r_P}.$$

The sum of distances from the center of force to the perihelion and aphelion (or to the perigee and apogee) equals the major axis $2a$ of the ellipse:

$$2a = r_A + r_P = \frac{2p}{1 - e^2}. \tag{10.16}$$

The sum of distances $r_1$ and $r_2$ from the foci to a point on the ellipse has the same value $2a$ for all points of the ellipse: $r_1 + r_2 = 2a$. This important property of the ellipse is often regarded as its definition. One of the simulation programs of the package `PLANETS AND SATELLITES` exploits this property in order to prove experimentally (by means of a computational experiment) the statement of Kepler's first law concerning the possible shapes of orbits in a Newtonian gravitational field.

One more property of the ellipse is worth mentioning: any light ray emerging from one of the foci of a concave elliptic mirror, after reflection, is directed toward the second focus of the ellipse. This *optical property* of the ellipse is related to Fermat's principle of geometrical optics. Fermat's principle states that the actual path of light between any two spatial points corresponds to a stationary value (in particular, minimum or maximum) of the optical length (and hence of the time of travel). For a plane mirror, this principle leads to the law of reflection: the angle of reflection equals the angle of incidence. For an ellipse, the sum of the distances from any point to the foci has the same value. Consequently, the optical path between the foci is the same for any light ray reflected by the elliptic mirror. Applied to a Keplerian orbit, this property means that, for any point of the orbit, the tangent (and thus the velocity vector) forms equal angles with the straight lines joining this point with the foci.

The distances between the center of the ellipse and its foci ($CF_1$ and $CF_2$ in figure 10.1) are equal to the product of the semimajor axis $a$ and the eccentricity $e$:

$$CF_1 = CF_2 = a - r_P = ae. \tag{10.17}$$

The semiminor axis $b$ is expressed in terms of the semimajor axis $a$ and eccentricity $e$ by:

$$b = a\sqrt{1 - e^2}. \tag{10.18}$$

## 10.5 Initial conditions and parameters of Keplerian orbits

Next we determine the parameters of a satellite's orbit when the satellite, whose initial distance from the center of a planet is $r_0$, has a transverse initial velocity $v_0$, as shown in figure 10.2. (By transverse, we mean that the velocity is perpendicular to the radius vector and hence to the local vertical line. Or, in other words, the initial velocity is horizontal.)

If the initial velocity equals the circular velocity $v_c$ for the given distance $r_0$, the orbit is a circle of radius $r_0$ (figure 10.2). If the initial velocity $v_0$ exceeds the circular velocity $v_c$, but is smaller than the escape velocity $v_{esc} = \sqrt{2}\,v_c$, the perigee $P$ of the elliptical orbit is located at the initial point and the apogee is at the opposite end $A$ of the major axis. This axis passes through the initial point and the center of the planet (see figure 10.2).

In order to find the distance $r_A$ between the center of force and the apogee, we can use the law of energy conservation and Kepler's second law or, equivalently, the law of conservation of angular momentum valid for an arbitrary motion in any central field (see section 11.1). At both the initial point $P$ and apogee $A$, the velocity vector is perpendicular to the radius vector $\mathbf{r}$ and the magnitude of the vector product of the

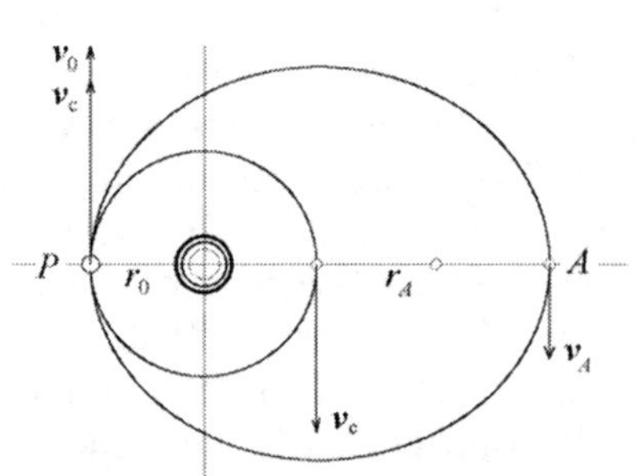


**Figure 10.2.** Circular and elliptical orbits of satellites that are launched from an initial point $P$ with transverse initial velocities.

velocity and radius vectors at these points equals the product of the magnitudes of **v** and **r** (since the sine of the angle between **v** and **r** equals 1):

$$v_0 r_0 = v_A r_A, \tag{10.19}$$

where $v_A$ is the velocity at the apogee. The second equation that is necessary for the determination of the two unknown quantities $v_A$ and $r_A$ is obtained by equating the values of the total energy at the initial point and at the apogee:

$$\frac{v_0^2}{2} - \frac{GM}{r_0} = \frac{v_A^2}{2} - \frac{GM}{r_A}. \tag{10.20}$$

Using (10.19), we then express the velocity $v_A$ at the apogee in terms of the initial velocity $v_0$ and the distances $r_0$ and $r_A$, and substitute it in the equation of the conservation of energy. Gathering the terms with $v_0$ on the left side of the equation, and moving the remaining terms to the right, we obtain:

$$v_0^2\left(1 - \frac{r_0^2}{r_A^2}\right) = \frac{GM}{r_0}\left(1 - \frac{r_0}{r_A}\right). \tag{10.21}$$

We can find the unknown distance from the center of the planet to the apogee, $r_A$, by solving this quadratic equation. One of its roots, $r_A = r_0$, corresponds to the initial point (to the perigee). This irrelevant root appears because the condition that we used for obtaining the equation, namely that the velocity vector be orthogonal to the radius vector, is satisfied also for the initial point (as well as for the apogee). In order to find the second root, the root that corresponds to the apogee, we express the difference of squares in the left side of the equation as the product of the corresponding sum and difference, and then divide both sides of the equation by $(1 - r_0/r_A)$. Then for the distance $r_A$ to the apogee of the orbit we obtain:

$$r_A = \frac{r_0}{2\left(v_c/v_0\right)^2 - 1}. \tag{10.22}$$

Here we have used (10.8) for the circular velocity $v_c$ at the initial distance $r_0$. The obtained expression is convenient for the determination of the parameters of the elliptical orbit in terms of the initial distance $r_0$ and the initial transverse velocity $v_0$. Figure 10.3 illustrates how the orbit depends on the initial velocity.

In the case $v_0 = v_c$ we have from (10.22) $r_A = r_0$, that is, the satellite moves in this case along a circular orbit whose radius equals $r_0$ (orbit *1* in figure 10.3).

The apogee distance $r_A$ increases as we increase the initial velocity (orbits *2–5* in figure 10.3). As the initial velocity approaches the value $v_{esc} = \sqrt{2}\,v_c$ of the escape velocity, a tiny increment in the initial velocity causes a large increment in the apogee distance.

At $v_0 = \sqrt{2}\,v_c$ the elliptical orbit elongates without limit and its apogee recedes to infinity. The ellipse becomes a parabola.

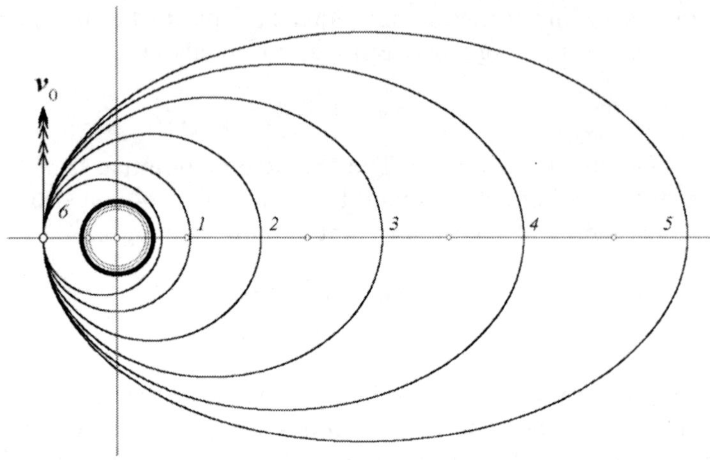


**Figure 10.3.** The orbits of satellites launched from the same initial point with different values for the initial velocities. Small circles indicate the second foci of the ellipses.

For values of the initial velocity greater than the escape velocity, the trajectory of the satellite is a hyperbola. In this case, (10.22) is not applicable because we assumed in its derivation that the motion is finite.

If the initial velocity $v_0$ is smaller than the circular velocity $v_c$ for the given initial distance $r_0$, we obtain from (10.22) for $r_A$ a value that is smaller than the initial distance $r_0$. This means that in this case the initial point is the apogee of the orbit, while the value $r_A$ given by (10.22) is the distance from the focus to the perigee, which in this case is located at the opposite end (with respect to the initial point) of the major axis passing through the initial point and the center of the Earth (orbit *6* in figure 10.3). Certainly, motion along the whole length of such an orbit is possible only if $r_A > R$, that is, if the perigee is outside the Earth so that the orbit does not intersect the Earth's surface.

The eccentricity of the elliptical orbit in terms of the transverse initial velocity, with the help of (10.22), is expressed as follows:

$$e = \frac{|r_A - r_0|}{r_A + r_0} = \left| \frac{v_0^2}{v_c^2} - 1 \right|. \tag{10.23}$$

The semimajor axis $a$ of the elliptical orbit is given by:

$$a = \frac{1}{2}(r_0 + r_A) = \frac{r_0}{2} \frac{1}{1 - v_0^2/(2v_c^2)}. \tag{10.24}$$

If the initial velocity equals the circular velocity, that is, if $v_0 = v_c$, this equation gives $a = r_0$, since the ellipse becomes a circle, and the semimajor axis coincides with the radius of the orbit. If $v_0 \to \sqrt{2}\, v_c$, that is, if the initial velocity approaches the escape velocity, (10.24) gives $a \to \infty$: the ellipse is elongated without limit. If $v_0 \to 0$, (10.24) gives $a \to r_0/2$: as the horizontal initial velocity becomes smaller and smaller, the elliptical orbit shrinks and degenerates into a straight segment connecting the initial point and the center of force. The foci of this degenerate, flattened ellipse are at the opposite ends of the line.

## 10.6 A satellite in the atmosphere

The air resistance experienced by a satellite in the rarefied strata of the upper atmosphere is caused by collisions between the satellite and air molecules. These collisions cause an exchange of energy and momentum between the satellite and air molecules, and lead to a gradual dissipation of the total mechanical energy of the satellite.

In the thermal equilibrium of the atmosphere, molecules of air move chaotically and continually collide with one another. If the mean free path covered by a molecule between successive collisions is much smaller than the linear dimensions of the satellite, we can treat the air as a continuous medium. When a body moves through a continuous medium, a *boundary layer* is formed around the surface of the body. Within this layer the velocities of the molecules of air are generally different from the velocities at large distances from the body, where the medium remains undisturbed. That is, the body partly involves in its motion the substance of the medium within the boundary layer.

The mean free path between collisions depends on the density of the gas. For the values of the density of air at the altitudes 160–200 km and higher, the mean free path is greater than the dimensions of a typical satellite by at least an order of magnitude. Thus no boundary layer is formed around the moving satellite. Any gas molecule, after an elastic collision with the satellite, flies unhindered so far from the satellite that its next collision most likely occurs with a molecule of air undisturbed by the satellite. In other words, we may regard air resistance at this altitude to be the result of *individual collisions* between the satellite and *separate molecules* of a macroscopically stationary gas rather than the ordinary friction experienced by a body that moves through a continuous medium. These separate molecules of air that the satellite meets on its way are in random thermal motion characterized by Maxwell's distribution of speeds of the undisturbed gas in the state of thermal equilibrium.

To support these conclusions, next we make simple numerical estimates. To evaluate the mean free path $\lambda$ of the molecule of air, we assume the molecule to be a sphere with a diameter $d \approx 3 \cdot 10^{-10}$ m. When the molecule moves through the distance $\lambda$ in the gas, on average it collides with another molecule only once. Thus in a cylinder of length $\lambda$ and cross-sectional area $\pi d^2$, on average there is only one molecule of the gas. Therefore,

$$n\pi d^2\lambda \approx 1, \tag{10.25}$$

where $n$ is the concentration, that is, the mean number of molecules per unit volume. We note that the mean free path $\lambda$ is determined only by the diameter of molecules and their concentration, and is independent of the velocity of molecules.

At an altitude of about 200 km, the density $\rho$ of air is approximately $10^{-9}$ kg m$^{-3}$. Since the mean mass of one mole of the air (mainly a mixture of nitrogen and oxygen) is $\mu = 0.029\,\text{kg mol}^{-1}$, we obtain a value $n = N_A\rho/\mu \approx 2 \cdot 10^{16}\ \text{m}^{-3}$ for the concentration of the air molecules (here $N_A = 6.02 \cdot 10^{23}\ \text{mol}^{-1}$ is Avogadro's number). With these values of concentration $n$ and molecular diameter $d$, (10.25) gives a value $\lambda \approx 200\,\text{m}$ for the mean free path. Since the mean free path of the air molecules in the upper atmosphere is much greater than the linear dimensions of an ordinary artificial satellite, we may consider air resistance to be caused by collisions of the satellite with individual air molecules.

The effect produced on a satellite by an immense number of collisions with the air molecules can be described on a macroscopic scale by introducing a resistive force that acts continuously on the satellite. This force depends on the speed of the satellite through the medium. This dependence is generally different for small and large speeds.

If a body moves slowly through the stationary air, so that the velocity $V$ of the body is much smaller than the average speed $\langle v\rangle$ of random thermal motion of the gas molecules, we can find the friction by using methods similar to those in the calculation of the pressure exerted by a gas on the wall of its container. Molecules striking the front of the body are more numerous and on average impart a greater momentum to the body in every impact than do molecules that are bombarding the body from the back. As a consequence, the pressure of the air exerted on the front of the moving body is greater than on the back. It can be shown from Maxwell's distribution of molecular velocities that in this case the force of friction is proportional to the *first power* of the velocity $V$ of the body.

On the other hand, we show below that if the velocity $V$ of the body is much greater than the mean thermal speed $\langle v\rangle$ of the gas molecules, air resistance is proportional to the *square* of the velocity $V$.

The mean thermal speed of the gas molecules $\langle v\rangle$ at atmospheric temperatures is approximately $500\,\text{m s}^{-1}$ and so is small compared with the orbital velocity $V$ of a satellite ($\approx 8\,\text{km s}^{-1}$). Hence we can ignore the thermal motion of the gas molecules in estimating the force of air resistance. In other words, we can assume that the satellite in its motion collides with stationary molecules of air. The interaction with the air takes place only on the front surface of the satellite.

For simplicity, we assume that collisions of the air molecules with the surface of the satellite are elastic. If the front surface is perpendicular to the velocity of the satellite, a momentum $2\,mV$ is transmitted to each molecule in a single collision. The number of such impacts per unit time equals the number of air molecules in a cylinder whose length equals the velocity $V$ of the satellite and whose base equals the area $A$ of the front surface (more generally, the cross-sectional area) of the satellite. Thus, the number of impacts per unit time equals the product of the concentration $n$ times the volume $A \cdot V$ of the cylinder.

The average force $F$ exerted by the satellite on the atmosphere is the total momentum imparted to the air molecules per unit time:

$$F = 2\,nmV^2A = 2\,\rho V^2A. \tag{10.26}$$

According to Newton's third law, an equal and opposite force is exerted on the satellite. This force of air resistance is proportional to the square of the velocity.

For example, if $\rho \approx 10^{-9}\,\mathrm{kg\,m^{-3}}$, $A \approx 1\,\mathrm{m^2}$, and the satellite moves with a velocity of $V \approx 10\,\mathrm{km\,s^{-1}}$, (10.26) yields that the force of air resistance equals approximately 0.2 N. Such a force, acting on a body of mass $M \approx 10^3$ kg, produces a deceleration of approximately $2 \cdot 10^{-4}\,\mathrm{m\,s^{-2}}$. (However, we remember that air resistance is not the only force acting on the satellite. It, in conjunction with the force of gravity, causes the satellite to speed up as its orbit decays in the upper atmosphere. This aerodynamical paradox of the satellite is discussed in section 4.2.)

The simplifying assumptions that the front surface of the satellite is perpendicular to the velocity, and that collisions of air molecules with the surface are elastic, are actually not very important for the validity of the above estimate of air resistance. In a completely inelastic collision, the momentum transmitted to the air molecule is half the value we have used above, and, consequently, the force of air resistance is somewhat smaller. In the case of completely inelastic collisions, the shape of the satellite does not influence the force; only its cross-sectional area is important. For a spherical satellite, it can be shown that the force of air resistance is the same both for completely elastic and inelastic collisions with air molecules.

Space-age flights have shown us that the atmosphere extends only to an altitude of about 100 km. This layer of air is very thin compared to the Earth's radius and its density decreases rapidly with altitude. The Earth's atmosphere is a gas placed in an open 'vessel without a cover'. Nothing keeps the air from escaping into outer space except the gravitational field of the Earth. This field determines the dependence of the air density on the altitude.

For a planet with a 'thin' atmosphere whose upper limit $h$ is much smaller than the planet's radius $R$ ($h \ll R$), we can assume the strength of the gravitational field $g$ of the planet to be approximately constant throughout the atmosphere. For an atmosphere that is held by a uniform gravitational field, the dependence of density on height in the state of thermal equilibrium obeys Boltzmann's distribution law:

$$\rho = \rho_0 \exp(-mgh/kT). \qquad (10.27)$$

Here $\rho_0$ is the density of air at sea level (at $h = 0$), $m$ is the mass of an average air molecule ($\approx$ 29 a.m.u. for the Earth's atmosphere), $g$ is the strength of the gravitational field near the surface of the planet, or the acceleration of free-fall ($9.8\,\mathrm{m\,s^{-2}}$ for the Earth), $k$ is Boltzmann's universal constant ($k = 1.38 \cdot 10^{-23}\,\mathrm{J\,K^{-1}}$), and $T$ is the absolute temperature. Equation (10.27) states that the density of air decreases *exponentially* with altitude.

Comparing (4.4) of chapter 2 and (10.27), we obtain for the *characteristic height* (or thickness) of the atmosphere the following expression:

$$H = kT/mg. \qquad (10.28)$$

For the Earth, substituting in (10.28) the corresponding values and assuming the equilibrium temperature of the atmosphere to be 25° C $\approx$ 300 K, we find $H = 8.8$ km. This value is very small compared with the Earth's radius $R$: $H/R \approx 0.0014$. This estimate shows that the approximation of a uniform gravitational field throughout the Earth's atmosphere is quite good. The assumption concerning the thermal equilibrium of the atmosphere is not so good. An improvement of this simplified model of the atmosphere is achieved if the characteristic height $H$ is considered as a function of the altitude $h$ because the air temperature $T$ changes (decreases) with altitude. For the altitudes $h < 120$ km the value of $H$ varies between 5 km and 10 km. Over the altitude of 120 km the air density decreases more slowly with height and the value of $H$ increases up to 30–40 km.

In the computer simulation, we can vary the value of $H$ widely. To make the effects of air resistance on the trajectories of satellites easily observable, we should enter exaggerated values for the height of the atmosphere.

## 10.7 Trajectories of a landing module

For a safe return to the Earth, a landing module should approach the dense strata of the atmosphere at a very small angle with the horizontal. A steep descent is dangerous because air resistance causes rapid heating of the module and, in the case of a manned spacecraft, because the astronauts may experience overloads of large g-factors. Therefore the descending trajectory should just graze the upper atmosphere.

We consider two possible impulse maneuvers to transfer the landing module from an initial circular orbit into a suitable descending trajectory: (a) the change in velocity is directed tangentially, antiparallel to the orbital velocity and (b) the change in velocity is directed radially, perpendicular to the orbital velocity.

In any case, an additional velocity transfers the space vehicle from the initial circular orbit to an elliptical orbit. One of the foci of the ellipse is located, in accordance with Kepler's first law, at the center of the Earth.

In case (a), the short-term impulse thrust of the rocket engine changes only the magnitude of the orbital velocity, preserving its direction. Therefore, at the point where the rocket engine operates (point $D$ in figure 10.4), the descending elliptical orbit has a common tangent with the original circular orbit. This point $D$ is the apogee of the elliptical orbit. Its perigee is located at the opposite end $P$ of the major axis, which passes through $D$ and the center of the Earth. At this point $P$ the ellipse should graze the dense strata of the atmosphere.

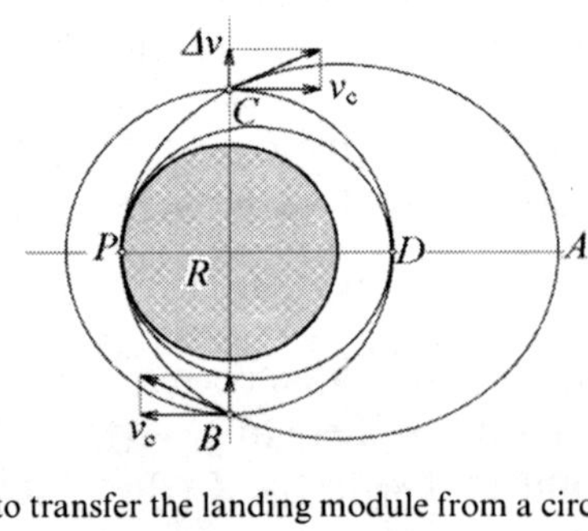


**Figure 10.4.** Possible maneuvers to transfer the landing module from a circular orbit to a trajectory grazing the planet.

To calculate the additional velocity $\Delta v$ (the *characteristic velocity*) that is necessary for the transition from the circular orbit to this descending elliptical trajectory, we make use of the conservation laws for energy and angular momentum.

We let $v_D = v_c - \Delta v$ be the velocity at the apogee $D$ of the elliptical orbit (here $v_c$ is the constant velocity in the original circular orbit), and $v_P$ be the velocity at the perigee $P$, where the ellipse grazes the globe. Then we write the laws of the conservation of energy and angular momentum for these points $D$ and $P$:

$$\frac{v_D^2}{2} - \frac{GM}{r_0} = \frac{v_P^2}{2} - \frac{GM}{R}, \qquad r_0 v_D = R v_P. \tag{10.29}$$

Here $r_0$ is the radius of the original circular orbit, $R$ is the Earth's radius (including the atmosphere) and $M$ is the mass of the Earth. Substituting $v_P$ from the second equation into the first, we obtain:

$$v_D^2\left(1 - \frac{r_0^2}{R^2}\right) = \frac{2GM}{r_0}\left(1 - \frac{r_0}{R}\right). \tag{10.30}$$

Dividing both parts of (10.30) by $(1 - r_0/R)$, we find the required value $v_D$ of the velocity at the apogee of the elliptical orbit:

$$v_D = \sqrt{\frac{2GM}{r_0}}\,\frac{1}{\sqrt{1 + r_0/R}}. \tag{10.31}$$

The first radial in the right side of (10.31) can be expressed in terms of the circular velocity $v_c$ for the original orbit: $v_c = \sqrt{GM/r_0}$. To find the value $\Delta v$ of the required change in velocity, we subtract $v_D$ from the circular velocity $v_c$:

$$\Delta v = v_c - v_D = v_c\left(1 - \sqrt{\frac{2}{1 + r_0/R}}\right). \tag{10.32}$$

For a low circular orbit, whose height $h = r_0 - R$ is much smaller than the Earth's radius $R$ (for $h \ll R$), the exact expression given by (10.32) can be transformed into a simpler and more convenient (though approximate) form. We substitute $r_0 = R + h$ into (10.32) and write the square root there as:

$$\sqrt{\frac{2}{1 + r_0/R}} = \sqrt{\frac{2}{2 + h/R}} = \frac{1}{\sqrt{1 + h/(2R)}} \approx 1 - \frac{h}{4R}.$$

Thus, from (10.32) we obtain the following approximate formula for calculation of the required additional (characteristic) velocity:

$$\Delta v = v_c\,\frac{h}{4R}. \tag{10.33}$$

For example, if the height $h$ of the circular orbit is $0.1\,R \approx 650$ km, the additional velocity $\Delta v$, according to (10.33), must be about 2.5% of the circular velocity.

(A calculation on the basis of (10.32) with $r = R + h = 1.1\ R$ gives a more exact value of 2.41%.)

In case (b), the additional velocity imparted to the space vehicle is directed radially, transverse to the orbital velocity, and both the magnitude and direction of the velocity change. Therefore the new elliptical orbit intersects the original circular orbit at point $B$ (see figure 10.4) at which the additional velocity $\Delta v$ is imparted to the landing module. For a soft landing, the new elliptical trajectory of descent must also graze the Earth (the upper atmosphere) at the perigee $P$ of the ellipse.

The laws of conservation of energy and angular momentum for points $B$ and $P$ in this case can be written as:

$$\frac{v_c^2 + (\Delta v)^2}{2} - \frac{GM}{r_0} = \frac{v_P^2}{2} - \frac{GM}{R}, \qquad v_c r_0 = v_P R. \tag{10.34}$$

Here the velocity $v_P$ at the perigee, as well as the additional velocity $\Delta v$, clearly have values different from those in (10.29). We note that the constant areal (sectorial) velocity in (10.34) for the descending elliptical trajectory has the same value as it does for the original circular orbit because an additional radial impulse from the rocket engine does not change the angular momentum of the landing module.

Substituting $v_P = v_c r_0 / R$ into the first equation (10.34) and taking into account that $GM/r_0 = 2v_c^2$, we obtain

$$v_c^2 + (\Delta v)^2 - 2v_c^2 = v_c^2\left(\frac{r_0}{R} - 1\right)^2.$$

Substituting $r = R + h$ in this equation, we finally obtain:

$$\Delta v = v_c \frac{h}{R}. \tag{10.35}$$

A comparison of (10.33) and (10.35) shows that for the second method, (b), the required additional velocity is approximately four times greater than that for the first method, (a). For example, it must equal 10% of the circular velocity if the height $h$ of the circular orbit is $0.1\ R$.

The angular distance between the starting point $B$ and the point of landing for method (b) is 90° (a quarter of the revolution), while for method (a) the angular distance between the point of transition from the circular orbit to the descending trajectory and the landing point is twice as large (180°, a half of the revolution).

In order to transfer the landing module to the trajectory that just grazes the Earth, using a transverse impulse, we can impart an additional velocity to the module not only downward but also vertically upward (see figure 10.4). It is clear from considerations of symmetry that in this case the required additional velocity $\Delta v$ has the same magnitude as it does for the downward impulse. However, to land at the same point $P$ on the Earth, the upward impulse must be imparted to the module at the opposite point of the original circular orbit (point $C$ in figure 10.4). The angular distance between point $C$ of the transition to the elliptical orbit and point $P$ of the landing in this case is 270° (three quarters of a revolution). The module at first rises higher. Then, only after it passes through the apogee of its elliptical orbit (point $A$ in figure 10.4), does it begin to descend toward the Earth's surface.

## 10.8 A space probe

In certain problems of space dynamics, the *relative motion* of the orbiting bodies is important. As an example, we consider a space probe launched from an orbital station in order to investigate the surface of some planet or interplanetary space. We assume that the station stays in a circular orbit around the planet. The probe (an automatic or manned space module with various scientific instruments) is to approach the surface of the planet and, after exploring the planet from a short distance and accumulating data, is to convey the data to the station.

To complete this mission successfully, a suitable elliptical orbit of the probe must have a sufficiently low perigee in order to approach the surface of the planet. Moreover, its period of revolution in its elliptical orbit must be commensurable with the period of the station: the space probe periodically meets with the station if their periods are in the ratio of small integers. If, for example, the periods of revolution of the probe and the station are in the ratio of 2 to 3, the probe after three revolutions returns to the common point of their orbits simultaneously with the station, just when the station has completed two revolutions. Examples of suitable inner orbits of the space probe are illustrated by figures 5.6 and 5.7 (see section 5.4).

Next we calculate the additional (characteristic) velocity that must be imparted to the space probe after its undocking from the station in order to transfer the probe to the elliptical orbit with the required period of revolution. If the additional velocity is directed tangentially to the circular orbit of the station, the semimajor axis $a$ of the elliptical orbit of the probe is determined by (10.24). With the help of this equation, we can express the square of the planetocentric initial velocity $v_0$ of the probe at the common point of the two orbits in terms of the semimajor axis $a$:

$$v_0^2 = v_c^2\left(2 - \frac{r_0}{a}\right). \tag{10.36}$$

This expression is valid both for the case of an initial velocity greater than the circular velocity, $v_0 > v_c$, and the case $v_0 < v_c$. The latter case corresponds to the additional velocity $\Delta v$ imparted to the space probe in the direction against the orbital velocity of the station. In this case the entire elliptical orbit of the probe lies inside the circular orbit of the station. The starting point at which the probe is undocked from the station is the apogee of such an inner-grazing elliptical orbit.

Next we can express in (10.36) the ratio $r_0/a$ in terms of the given ratio of the period $T_0$ of the station to the period $T$ of the probe in its elliptical orbit with the semimajor axis $a$. We do this with the help of Kepler's third law:

$$\frac{r_0}{a} = \left(\frac{T_0}{T}\right)^{2/3}. \tag{10.37}$$

**Table 10.1.** Inner orbits of the space probe.

| $T_0/T$ | $v_0/v_c$ | $\Delta v/v_c$ | $r_P/r_0$ |
|---|---|---|---|
| 2/1 | 0.642 34 | 0.357 66 | 0.259 92 |
| 3/2 | 0.830 50 | 0.169 56 | 0.526 29 |
| 4/3 | 0.888 02 | 0.111 98 | 0.650 96 |
| 5/4 | 0.916 30 | 0.083 70 | 0.723 55 |

**Table 10.2.** Outer orbits of the space probe.

| $T_0/T$ | $v_0/v_c$ | $\Delta v/v_c$ | $r_A/r_0$ |
|---|---|---|---|
| 4/5 | 1.066 876 | 0.066 88 | 1.320 79 |
| 3/4 | 1.083 752 | 0.083 75 | 1.422 82 |
| 2/3 | 1.112 140 | 0.112 14 | 1.620 74 |
| 1/2 | 1.170 487 | 0.170 49 | 2.174 80 |

Hence, after the undocking, the space probe must have the following planetocentric velocity:

$$v_0 = v_c\sqrt{2 - \left(T_0/T\right)^{2/3}}. \qquad (10.38)$$

Table 10.1 lists the values of the initial velocity $v_0$ of the space probe and the corresponding values of the additional (characteristic) velocity $\Delta v = |v_0 - v_c|$ for several inner elliptical orbits of the probe. (These velocities are expressed in units of the circular velocity $v_0$ of the orbital station for convenience of use in the simulations.) The perigee distance $r_P = 2a - r_0$ for each of the orbits (in units of the radius $r_0$ of the station's circular orbit) is also listed. An orbit is possible if this distance is greater than the radius $R$ of the planet. The difference $r_P - R$ is the minimal distance from the surface of the planet reached by the space probe.

If the additional velocity imparted to the probe is directed forward, tangentially to the orbit of the station, the resulting elliptical orbit encloses (circumscribes) the circular orbit of the station. The initial point at which the orbits graze one another is the perigee of the elliptical orbit. Such outer orbits of space probes with suitable periods of revolution may be used to investigate interplanetary space. Examples of the outer orbits are illustrated by figure 5.8 of section 5.4.

Table 10.2 lists the values of the initial velocity $v_0$ of the space probe and the corresponding values of the additional velocity $\Delta v = v_0 - v_c$ for several outer elliptical orbits of the probe. The apogee distance $r_A = 2a - r_0$ for each of the orbits (the greatest distance of the probe from the center of the planet) is also listed.

If the additional velocity $\Delta v$ is imparted to the probe in the radial direction (vertically up or down), the period of revolution is always greater than the period of the orbital station. Applying to this case the laws of the conservation of energy and angular momentum (Kepler's second law), and taking into account that a radial impulse of the rocket thrust does not change the angular momentum of the space probe, we can obtain the following expressions for the distances of the apogee and perigee of the elliptical orbit from the center of the planet:

$$r_A = \frac{r_0}{1 - \Delta v / v_c}, \qquad r_P = \frac{r_0}{1 + \Delta v / v_c}. \tag{10.39}$$

Therefore, the semimajor axis of the elliptical orbit of the space probe depends on the magnitude $\Delta v$ of the transverse additional velocity as follows:

$$a = \frac{1}{2}(r_A + r_P) = \frac{r_0}{1 - \left(\Delta v / v_c\right)^2}. \tag{10.40}$$

Suitable orbits for the space probe must have certain periods of revolution. We can use Kepler's third law $r_0/a = \left(T_0/T\right)^{2/3}$ to calculate the additional velocity $\Delta v$ that gives an orbit with the required period of revolution $T$. With the help of (10.40), we obtain

$$\left(\frac{\Delta v}{v_c}\right) = 1 - \left(\frac{T_0}{T}\right)^{2/3}. \tag{10.41}$$

For example, to obtain the orbit of the space probe with a period that is one and a half periods of the orbital station ($T_0/T = 2/3$), the required additional velocity $\Delta v$ calculated from (10.41) is 0.486 68 $v_c$. Such a probe returns to the station after every two revolutions in its elliptical orbit. During this time the station makes three revolutions in its circular orbit. Such an elliptical orbit is shown in figure 5.9 of section 5.4.

For $T_0/T = 4/5$ (10.41) gives $\Delta v / v_c = 0.37179$. In this case the space probe and the station meet after every four revolutions of the probe and five revolutions of the orbital station.

## 10.9 Space rendezvous

The laws of the conservation of energy and angular momentum, together with Kepler's laws of motion in a central Newtonian gravitational field, can be used in calculating the maneuvers required for a planned space flight between two circular orbits and for an approximate calculation of an interplanetary flight.

Next we consider a semielliptic Hohman's transition between two circular orbits. We assume for definiteness that we wish to launch a spacecraft from an orbital station that moves around a planet in a circular orbit of radius $r_0$ into an outer circular orbit of radius, say, $2r_0$. After the spacecraft remains for some time in this new orbit, it is to return to the station and dock to it. The simulation experiment for such maneuvers is described in section 5.5 (see figure 5.10). Here we present the calculations for the required characteristic velocity and for the time moments at which the maneuvers take place.

The ellipse of the semielliptic transitional trajectory that ensures the most economical transition (in terms of expending rocket fuel) grazes both the initial circular orbit (from the outside) and the final circular orbit (from the inside). Hence the perigee distance from the center of the planet equals $r_0$, the radius of the initial orbit, and the apogee distance equals $2r_0$, the radius of the final circular orbit. To calculate the velocity $v_0$ that the spacecraft must have at perigee of the semielliptic transitional trajectory, we can use (10.22):

$$v_0 = v_c\sqrt{\frac{2}{1 + r_0/r_A}}. \qquad (10.42)$$

Here $r_A$ is the apogee distance of the elliptical orbit from the center of the planet. To find the required additional velocity $\Delta v_1$ for the first maneuver, we subtract from $v_0$ (10.42), the circular velocity $v_c$ which the spacecraft already has after undocking from the station:

$$\Delta v_1 = v_c\left(\sqrt{\frac{2}{1 + r_0/r_A}} - 1\right). \qquad (10.43)$$

Substituting $r_A = 2r_0$, we obtain from (10.43) $\Delta v_1/v_c = 2/\sqrt{3} - 1 = 0.1547$.

The spacecraft comes to the apogee with a velocity $v_A$, whose value is related to the velocity $v_0$ at the perigee, (10.42), through the law of the conservation of angular momentum (Kepler's second law):

$$v_0 r_0 = v_A r_A.$$

For $r_A = 2r_0$ we find, with the help of (10.42), $v_A = v_0/2 = 0.577\, v_c$. To transfer the spacecraft from the elliptical orbit to the circular orbit of radius $2r_0$, we must increase the velocity at the apogee by a second jet impulse. The circular velocity in a given central Newtonian gravitational field is inversely proportional to the square root of the radius of the circular orbit. For the orbit of radius $2r_0$, the circular velocity equals $v_c/\sqrt{2} = 0.707\, v_c$, where $v_c$ is the circular velocity for the original orbit of radius $r_0$. Subtracting from this value the velocity $v_A = 0.577\, v_c$, with which the spacecraft reaches the apogee of the elliptical orbit, we find the additional velocity $\Delta v_2$ required for the second maneuver: $\Delta v_2/v_c = 0.707 - 0.577 = 0.130$.

Next we calculate the time moments at which these maneuvers take place. We can do so with the help of Kepler's third law. The semimajor axis $a$ of the elliptical orbit equals $(r_0 + r_A)/2 = (3/2)r_0$. We call the period of revolution along the original circular orbit (orbit of the station) $T_0$. Then the period for the elliptical orbit equals $(a/r_0)^{3/2}\, T_0 = 1.5^{3/2}\, T_0 = 1.837\, T_0$. If we assume $t_1 = 0$ for the first jet impulse, the second jet impulse must be imparted to the spacecraft after a lapse of one half of the period for the elliptical orbit, that is, at $t_2 = 0.9186\, T_0$.

During the period of time $t = t_2 - t_1$ taken for the transition, the radius vector of the station rotates through an angle of $(2\pi/T_0)t$ radians. Since the radius vector of the spacecraft turns during this semielliptic transition through the angle $\pi$, at the instant of the second maneuver the spacecraft lags behind the station by an angle $\alpha = (2\pi/T_0)t - \pi = 2\pi(0.9186 - 0.5) = 2\pi \cdot 0.4186$ radians.

After the spacecraft remains in its new circular orbit for a while, it is to return to the orbital station. The optimal return path between the two circular orbits is again semielliptic. The additional velocity $\Delta v_3$ in the jet impulse that transfers the spacecraft from the outer orbit to the semielliptic transitional trajectory is directed against the orbital velocity. It is clear from symmetry that in magnitude the additional velocity this time must be exactly the same as for the preceding transition from the elliptical trajectory to the outer circular orbit, that is, $\Delta v_3 = \Delta v_2 = 0.130\ v_c$. And when the spacecraft reaches the perigee of the elliptical trajectory where it grazes the inner circular orbit, one more jet impulse is necessary to quench the excess velocity. This time the additional velocity $\Delta v_4$ must have the same magnitude as it did for the first transition from the initial circular orbit to the semielliptic trajectory: $\Delta v_4 = \Delta v_1 = 0.1547\ v_c$.

However, the return journey of the spacecraft is complicated by the fact that it is not sufficient to simply transfer the spacecraft to the original inner circular orbit. The spacecraft must reach the grazing point of the transitional semielliptic trajectory and the inner circular orbit just at the moment when the orbital station arrives at this point. To ensure the rendezvous, we must choose a proper moment for the transition from the outer orbit to the semielliptic return path. What should the system configuration be at this moment?

During the direct transition to the outer orbit, the *spacecraft lagged* behind the station by an angle of $\alpha = 2\pi \cdot 0.4186$ radians ($\alpha$ is the angle between the radius vectors of the station and the spacecraft at $t = t_2$). The journey back takes place over the same duration of time as does the journey out. Consequently, in order to meet with the station, the spacecraft must begin its journey back at the moment when the *station is behind* the spacecraft by the same angle $\alpha$.

Letting $T$ be the period of revolution of the spacecraft along the outer circular orbit, it follows from Kepler's third law that $T = 2^{3/2}\ T_0 = 2.83\ T_0$, since the radius of the outer orbit is $2r_0$. Calling $\Delta\omega$ the difference between the angular velocity $2\pi/T_0$ of the station and the angular velocity $2\pi/T$ of the spacecraft, we have that $\Delta\omega = (2\pi/T_0) \cdot 0.646$. The angular distance $\beta(t)$ between the station and the spacecraft at an arbitrary time $t > t_2$ is determined by the expression:

$$\beta(t) = \Delta\omega(t - t_2) + \alpha, \qquad (10.44)$$

since at $t = t_2$ this angular distance equals $\alpha$. To calculate the time $t_3$ suitable for starting the return journey, we require that at that moment the station be behind the spacecraft by $\alpha$. Consequently, the angle $\beta$ given by (10.44) should be made equal to $2\pi n - \alpha$, where $n$ is an integer:

$$2\pi n - \alpha = \Delta\omega(t_3 - t_2) + \alpha. \qquad (10.45)$$

Since $\alpha = 2\pi \cdot 0.4186$ radians, we find from (10.45) that the time $t_3 - t_2$ during which we can stay in the outer circular orbit is given by:

$$t_3 - t_2 = T_0(n - 0.8372)/0.646. \tag{10.46}$$

For $n=1$ (10.46) gives $t_3 - t_2 = 0.252\ T_0$. During this interval the spacecraft covers only a small part of the outer orbit. And so if the spacecraft is to remain longer, we let $n=2$ in (10.46) to find that $t_3 - t_2 = 1.7987\ T_0$. The period of revolution for the outer circular orbit equals $2^{3/2}\ T_0 = 2.83\ T_0$, and so with $n=2$ the spacecraft covers a considerable part of the orbit. If we are satisfied with this duration, the third maneuver must be performed at $t_3 = t_2 + 1.7987\ T_0 = 2.7174\ T_0$. Adding the duration $0.9186\ T_0$ of motion along the semielliptic trajectory, we find the moment $t_4$ at which the rendezvous of the spacecraft with the station occurs: $t_4 = 3.636\ T_0$. At this moment the fourth jet impulse of a magnitude $\Delta v_4 = \Delta v_1 = 0.1547\ v_c$ must be imparted to the spacecraft in order to equalize its velocity with the orbital velocity of the station.

The above discussion illustrates how space maneuvers are calculated using Kepler's laws and the laws of conservation of energy and angular momentum. These calculations can be tested using the simulation programs described in section 5.5. Figure 5.10 illustrates the particular maneuvers calculated above.

## 10.10 Kepler's laws and the solar system

As mentioned above, the true shape of planetary orbits was discovered through trial and error by Johannes Kepler in about 1610, from the careful, laborious and prolonged astronomical observations of Tycho Brahe. According to Kepler's first law, the orbits of planets are ellipses with the Sun at one focus of the ellipse. Although the law was originally established for planetary motion, it is also valid for any motion under a central force of attraction that decreases as the square of the distance from the center of force. In particular, the free motion of natural and artificial satellites around the Earth and other planets also obeys Kepler's first law—orbits of satellites are ellipses or circles.

The orbit of a celestial body is exactly an ellipse in the idealized case of its motion solely under the action of a central inverse-square force. The dynamical explanation of Kepler's first law was first given by Isaac Newton on the basis of his laws of motion (Newton's second law) and the law of universal gravitation. Possible analytic derivations of Kepler's first law are given in sections 11.2 and 11.5 of this textbook.

For the planets of our Solar System, Kepler's first law is a good zeroth-order approximation because the masses of planets are small compared to the mass of the Sun and the planets are separated from one another by large distances. That is, with good precision we can neglect the forces of gravitation between the planets and consider their motion to be governed only by their attraction to the Sun. Hence, because of the structure of our Solar System, the motion of each of the planets is rather simple.

Many stars in our Galaxy are multiple systems—double and triple stars, unlike the Sun, which is a single star. However, stable planetary orbits are also possible in a multiple star system and it is conceivable that a community of animated, thinking creatures could arise on such a planet. Because the trajectories of planets in a double-star system are very complicated, it would be an immensely difficult problem for astronomers among those creatures to establish the kinematic laws of planetary motion in the double-star system and an even more difficult problem would be to discover that these complex kinematical laws are generated by the simple inverse-square law of gravitational attraction to each of the stars. Our civilization must be grateful to whatever powers that may control its destiny for the happy circumstance of inhabiting a planet that orbits a single star. Humankind has been lucky to travel so fast along the thorny road of knowledge.

One of the simulation programs of the package `PLANETS AND SATELLITES` deals with a planet orbiting a double star. If the mass of the planet is small compared with the masses of the primaries, the influence of the planet on them is negligible. Such a system is an example of the restricted three-body problem. Although the components of the double star execute rather simple periodic Keplerian motions around the center of mass of the system, the motion of the planet may be very complicated. In the general case, it is impossible to obtain an analytic solution to the problem. The absence of such a solution probably reflects the complexity of the possible motions of the system rather than the weakness of the analytic capability of the mathematics. The simulation programs enable us to experiment with various external planetary orbits (those which encompass both stars) and internal orbits (those encompassing only one of the stars). The simulations display just how complicated the motion of such a planet can be, in spite of the simplicity of the fundamental laws of physics that govern this motion.

## 10.11 An approximate approach to the restricted three-body problem

The motion of planets around the Sun is almost entirely governed by their attraction to the Sun. The masses of the planets of the Solar System are small compared to the mass of the Sun and so the gravitational forces between the planets cause rather small deviations from Kepler's laws. In the case of a single star with a single orbiting planet, the motion is exactly Keplerian.

But what can we say about the motion of a satellite orbiting a planet? To what extent can we consider this motion to be Keplerian? Besides the gravitational attraction to the planet, the satellite is subjected to the gravitational attraction to the Sun. For example, a simple calculation shows that the force of gravitational attraction to the Sun of our Moon is greater than the force of its attraction to the Earth. Does it make sense to say that the Moon orbits the Earth in spite of the fact that the attraction of the Moon to the Sun is greater than to the Earth?

To answer this question, we should remember that in our description the motion of a satellite is referred to a reference frame associated with the planet rather than with the Sun. However, the reference frame associated with the planet is not an inertial one. Together with the planet, it is subjected to the acceleration directed toward the Sun. When the satellite is not far from the planet, the gravitational pull of the Sun gives the satellite almost the same acceleration as the Sun gives to the planet itself. Indeed, in a uniform gravitational field, all bodies (in our case the satellite and the planet) have equal accelerations independent of their masses. This follows from the equivalence of gravitational and inertial masses. The force of gravity exerted on a body is proportional to its gravitational mass, while the acceleration produced by a force is inversely proportional to its inertial mass[2].

Hence the influence of the Sun on the motion of satellites relative to the planet is not very significant and so in its principal features this motion is described by Kepler's laws. The gravitational attraction of a satellite toward the Sun reveals itself only as a perturbation in the acceleration of the satellite. This perturbation is equal not to the acceleration produced by the attraction of the satellite to the Sun, but rather to the difference of the accelerations of the satellite and the planet, produced by the gravitational field of the Sun. Since these accelerations are almost equal, their difference is small compared to the acceleration of the satellite produced by its gravitational attraction to the planet.

In other words, the Keplerian orbital motion of the satellite around the planet is actually perturbed not by the gravitational field of the Sun by itself, but rather by the *non-uniformity* of this field. In our everyday life on the Earth, this non-uniformity of the gravitational field of the Sun, as well as the non-uniformity of the gravitational field of the Moon, reveals itself in the ocean tides. The tidal force (the differential gravitational force), in contrast to the total gravitational force, decreases as the cube, not the square, of the distance between two bodies.

Because the orbits of the artificial satellites of the Earth are relatively small, the non-uniformity of the gravitational field of the Sun is rather insignificant. Therefore, calculating the motion of a satellite relative to the planet, we can consider its gravitational attraction only by the planet as a first approximation. In other words, the motion of a satellite around a planet can be analytically investigated to the first approximation as a restricted two-body problem, which has an exact solution. For a satellite whose mass is negligible compared to the mass of the planet, this relative motion is simply a Keplerian motion in a Newtonian central gravitational field. The orbit of a satellite around the planet can be regarded to the first approximation as an ellipse or a circle, and the infinite trajectory of a spacecraft bypassing a planet can be regarded, in the neighborhood of the planet, as a segment of a parabola or a hyperbola.

The complicated looping trajectory of a satellite in the heliocentric frame of reference is explained by the addition of two simple motions: motion along a large circle (or an ellipse) in which the satellite follows the planet around the Sun and the simultaneous revolution of the satellite about the planet along a much smaller circle (or smaller ellipse).

---

[2] Or, vice versa, one may treat the equality of accelerations acquired by all bodies in a given gravitational field (reliably established experimentally with great accuracy) as experimental evidence for the equivalence of the inertial and gravitational masses.

We call the exact Keplerian motion of a satellite relative to a planet unperturbed motion. The motion of satellites in low orbits that pass not far from the planet can be considered as an unperturbed Keplerian motion with rather good precision. However, for large distances of a satellite from the planet, the perturbing influence caused by the non-uniformity of the gravitational field of the Sun on the motion of the satellite around the planet becomes important.

Therefore a problem arises concerning the determination of the region around the planet within which the motion of a satellite can be considered as unperturbed. Such a region is called the *sphere of gravitational action* of the planet with respect to the Sun. A definition of the sphere of gravitational action and a derivation of its radius is found in section 11.2.

The concept of the sphere of gravitational action of a smaller celestial body with respect to a heavier one proves to be very helpful in obtaining approximate solutions to the *restricted three-body problem.* The calculation of the trajectory for a space expedition from the Earth to the Moon gives an example of such a problem.

In the restricted three-body problem, the motion of the light body (e.g. a spacecraft whose mass is assumed to be zero) is the most interesting. This motion occurs under the forces of gravitation created by the two massive bodies whose motion can be considered as known. As we already mentioned, even the restricted three-body problem does not have a general analytic solution.

An approximate solution to the restricted three-body problem can be obtained by the *method of joined conic sections*. The principal idea of the method is to ignore the influence of the second massive celestial body on the motion of the spacecraft until the latter enters the sphere of gravitational action of this body. In other words, we consider the motion in the frame of reference associated with the larger massive body as an unperturbed Keplerian motion in the gravitational field of this body. After the spacecraft enters the sphere of gravitational action of the smaller celestial body, we consider its motion in the non-inertial reference frame associated with the smaller body and assume that this motion is governed solely by the gravitational field of the smaller body.

On the boundary of the sphere of gravitational action we join the two Keplerian orbits. That is, we transform the coordinates and velocity of the spacecraft from one frame of reference to the other and regard these new values as the initial conditions for the continued Keplerian motion in the new frame of reference. Thus the three-body problem is reduced to two two-body problems, for which exact analytic solutions are possible.

Clearly there are no boundaries in space that are impenetrable to gravitational fields. The division of space into separate regions in which the motion of a spacecraft is governed solely by one celestial body is no more than a convenient convention. Therefore this method for obtaining a solution to the restricted three-body problem is an approximation. The verification of these approximate solutions by the direct numerical integration of the equations of motion shows that the method of joined conic sections gives not only good qualitative results but also rather reliable quantitative estimates.

## Questions for further thought

1. Why can celestial bodies such as planets and their moons only exist in motion? Why can a planetary system with immovable planets not exist? Why can there not be even a single immovable body (a planet or moon) in a planetary system or even in a galaxy? Why then do we perceive the stars in the sky as immovable?
2. Which physical laws give an explanation to the actual (circular or elliptic) shapes of planetary orbits? Which peculiarity of the law of universal gravitation is responsible for the existence of closed orbits? Specify the requirements (physical conditions) under which the orbit of a celestial body would be exactly elliptical.
3. Kepler's first law claims that a body under the central inverse-square gravitational force traces an elliptical orbit. What are the properties of these motions that are described by Kepler's second and third laws?
4. Which peculiarities in the construction of the Solar System are responsible for the (approximately) elliptical shape of planetary orbits? Why are the real orbits only approximately elliptical?
5. The force of gravitational attraction to the Sun of our natural satellite, the Moon, is greater than the gravitational force exerted on the Moon by the Earth. Why then does it make sense to say that the Moon orbits the Earth? What does the Moon's orbit looks like in the geocentric (associated with the Earth) and heliocentric frames of reference?
6. What are the physical reasons that explain the (almost) constant tilt (inclination) of the axis of Earth's daily rotation to its orbit (this inclination equals 23.5° and is responsible for the change of seasons on the Earth)?
7. Although the tilt of the Earth's axis during the millennia remains almost the same, the orientation of the Earth's axis changes slowly in space—the axis moves along a cone with a period of almost 26 000 years. Such a motion is called precession and was the third-discovered motion of the Earth, after the far more obvious daily rotation and annual revolution. Because of precession, the north celestial pole, which today points to within 1° of the arc of Polaris (and will point closest to Polaris in 2017), will, in 12 000 years, point about 5° from Vega. What are the physical reasons for this precession?

**IOP** Publishing

Motions of Celestial Bodies
Computer simulations
**Eugene Butikov**

# Chapter 11

## Theoretical background

This chapter is intended for a much more sophisticated and in-depth study of the subject. This highly mathematical chapter provides rigorous derivations and delves into the serious theoretical background for the computer-aided study of celestial mechanics and astrodynamics.

### 11.1 Angular momentum and areal velocity

The *angular momentum* of a point particle of mass $m$ with respect to the origin is defined as the vector product of the radius vector $\mathbf{r}$ to the particle and the vector of the *linear momentum* $\mathbf{p} = m\mathbf{v}$, where $\mathbf{v}$ is the velocity of the particle:

$$\mathbf{L} = \mathbf{r} \times \mathbf{p} = \mathbf{r} \times m\mathbf{v}. \tag{11.1}$$

We can find the rate of change of the angular momentum in time $\mathrm{d}\mathbf{L}/\mathrm{d}t$ by taking the time derivative of the right side of (11.1), considering it as a product of two factors $\mathbf{r}$ and $m\mathbf{v}$:

$$\frac{\mathrm{d}\mathbf{L}}{\mathrm{d}t} = \frac{\mathrm{d}\mathbf{r}}{\mathrm{d}t} \times m\mathbf{v} + \mathbf{r} \times m\frac{\mathrm{d}\mathbf{v}}{\mathrm{d}t}. \tag{11.2}$$

The first term in the right-hand side of (11.2) is zero because the derivative $\mathrm{d}\mathbf{r}/\mathrm{d}t$ is the velocity $\mathbf{v}$ of the particle and the vector product of a vector with itself is zero: $\mathbf{v} \times \mathbf{v} = 0$. In the second term, the acceleration $\mathbf{a} = \mathrm{d}\mathbf{v}/\mathrm{d}t$ can be expressed in terms of the net force $\mathbf{F}$ on the particle and its mass. From Newton's second law:

$$m\frac{\mathrm{d}\mathbf{v}}{\mathrm{d}t} = \mathbf{F}. \tag{11.3}$$

 

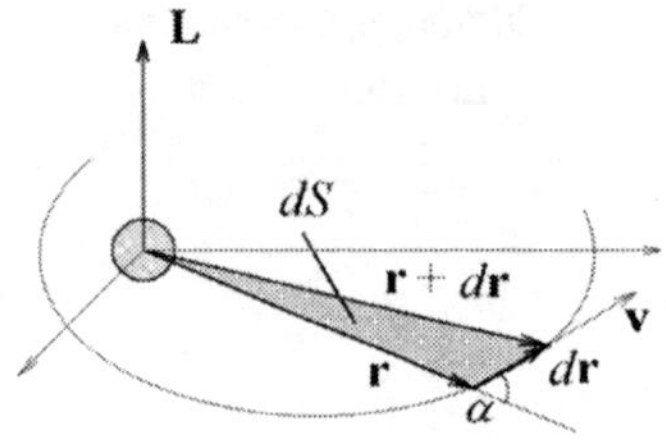


**Figure 11.1.** The geometric meaning of angular momentum.

Thus we obtain that the time rate of change of the angular momentum equals the torque of the net force **F** on the particle relative to the origin:

$$\frac{\mathrm{d}\mathbf{L}}{\mathrm{d}t} = \mathbf{r} \times \mathbf{F}. \tag{11.4}$$

By definition, when a particle moves in a central field, the force **F** of the field is directed along the radius vector **r** and the torque of the force relative to the center of the force is thus zero: $\mathbf{r} \times \mathbf{F} = 0$. Consequently, as we can see from (11.4), the angular momentum of a particle relative the center of force is conserved during the motion: $\mathbf{L} = \text{const}$. As we show next, this conservation of the angular momentum implies the constancy of the areal velocity. In other words, Kepler's second law is a consequence of the conservation of angular momentum.

Let us consider the geometric meaning of the angular momentum of a particle orbiting a central body (figure 11.1). We can represent velocity **v** in the definition of the angular momentum, (11.1), as the ratio of the vector d**r** of infinitesimal displacement to the corresponding infinitesimal time interval d$t$:

$$\mathbf{L} = \mathbf{r} \times m\mathbf{v} = m\mathbf{r} \times \mathrm{d}\mathbf{r}/\mathrm{d}t. \tag{11.5}$$

The vector product $\mathbf{r} \times \mathrm{d}\mathbf{r}$ in the right side of (11.5) is a vector perpendicular to the plane in which the vectors **r** and d**r** lie (that is, the vectors **r** and **v**). The magnitude of this vector product,

$$|\mathbf{r} \times \mathrm{d}\mathbf{r}| = r\mathrm{d}r \sin\alpha = 2\mathrm{d}S, \tag{11.6}$$

is twice the area d$S$ of the elementary triangle swept out by **r** during d$t$. (This area is shaded in figure 11.1.) Indeed, the product of d$r$ with the sine of the angle $\alpha$ between the vectors **r** and d**r** is the height of this triangle (measured from its side **r**). The ratio d$S$/d$t$ of the elementary area d$S$ to the time interval d$t$ during which the radius vector d**r** 'sweeps out' this area, is called the *areal velocity* or *sectorial velocity*. Therefore, it follows from (11.6) that the magnitude of the angular momentum is proportional to the areal velocity:

$$L = 2m\frac{\mathrm{d}S}{\mathrm{d}t}. \tag{11.7}$$

The conservation of the *direction* of the angular momentum vector during the motion means that the trajectory of a particle in a central field is a plane curve, that is, it lies in a fixed plane orthogonal to the constant vector **L**. The orientation of this plane is determined by the initial values of the radius vector $\mathbf{r}_0$ and velocity $\mathbf{v}_0$. The conservation of the *magnitude* of the angular momentum means that the areal velocity is constant.

Thus, Kepler's second law is a consequence of the conservation of angular momentum for a particle in a central field. The areal velocity is constant for all Keplerian motions, including motions along parabolic and hyperbolic trajectories, as well as motions along elliptical (and circular) orbits. We emphasize that Kepler's second law holds for *any* central field, not only for the inverse-square gravitational field. On the other hand, Kepler's first and third laws are valid only for the motions in a Newtonian (or Coulomb) central field, whose force is inversely proportional to the square of the distance from the center of force.

## 11.2 Dynamical derivation of Kepler's first law

It follows mathematically from Newton's second law that regular planetary motions under the inverse-square force of attraction to the Sun are described by Kepler's laws. All the planets (and the asteroids) move under this force around the Sun in elliptical orbits. Kepler's laws are derivable from Newton's laws of motion with a central force of gravity varying as $1/r^2$ from a fixed point and, vice versa, Newton's law of gravity is derivable from Kepler's laws if one assumes Newton's laws of motion.

The simulation program that illustrates Kepler's laws can be regarded as their experimental verification (in a computation experiment) because the program actually displays the motion without 'knowing' anything about these laws: the program exploits only Newton's second law of motion and the law of universal gravitation.

Next we prove analytically that trajectories in a Newtonian (inverse-square) gravitational field are *conic sections*. It is convenient to use the laws of conservation of angular momentum and energy for the derivation of the shape of the trajectory rather than Newton's second law directly. We use the polar coordinates $r$ and $\varphi$ to indicate the position of the particle in the plane of motion. Let the origin of the coordinate system be at the center of force.

The equation of a trajectory in polar coordinates is $r = r(\varphi)$, an equation that expresses the distance $r$ from the center as a function of the angle $\varphi$ between the radius vector and some fixed direction (the polar axis) in the plane.

First we express the magnitude of the angular momentum $L$ of a particle in terms of its polar coordinates:

$$L = m|\mathbf{r} \times \mathbf{v}| = mrv_\perp = mr^2\dot{\varphi}. \tag{11.8}$$

Here $v_\perp = r\dot{\varphi}$ is the transverse component of the particle's velocity (the component orthogonal to the radius vector). Since the angular momentum $L$ of the particle remains constant during motion in a central field, for any point of the trajectory the angular velocity $\dot{\varphi}$ can be expressed (using (11.8)) in terms of the distance $r$ from the origin (from the center of force) and the constant value of angular momentum $L$:

$$\dot{\varphi} = \frac{L}{mr^2}. \tag{11.9}$$

Next we use the conservation of energy. In the expression $mv^2/2$ for the kinetic energy, the square of the particle's velocity is the sum of the squares of its radial ($\dot{r}$) and transverse ($r\dot{\varphi}$) components: $v^2 = \dot{r}^2 + r^2\dot{\varphi}^2$. Substituting $\dot{\varphi}$ from (11.9) into the second term, we write for the total energy $E_{\text{kin}} + U$ in polar coordinates:

$$\frac{1}{2}m\dot{r}^2 + \frac{L^2}{2mr^2} - G\frac{mM}{r} = E. \tag{11.10}$$

The constant values $E$ of the total energy and $L$ of the angular momentum in (11.10) are determined by the initial conditions. To find the shape of the trajectory $r = r(\varphi)$, we eliminate time from (11.10). Considering $r$ as a function of $\varphi$ rather than of $t$ explicitly, we write:

$$\dot{r} = \frac{\mathrm{d}r}{\mathrm{d}t} = \frac{\mathrm{d}r}{\mathrm{d}\varphi}\frac{\mathrm{d}\varphi}{\mathrm{d}t} = \frac{\mathrm{d}r}{\mathrm{d}\varphi}\frac{L}{mr^2}. \tag{11.11}$$

Here we have expressed the angular velocity $\mathrm{d}\varphi/\mathrm{d}t = \dot{\varphi}$ in terms of the angular momentum $L$ with the help of (11.9). Substituting this expression for $\dot{r}$ into equation (11.10), we obtain a differential equation for the function $r(\varphi)$ that describes the trajectory.

This differential equation can be simplified if we introduce a new function $\rho = \rho(\varphi)$ instead of $r(\varphi)$ by the relation $\rho = 1/r$. Since

$$\frac{\mathrm{d}r}{\mathrm{d}\varphi} = \frac{\mathrm{d}}{\mathrm{d}\varphi}\frac{1}{\rho} = -\frac{1}{\rho^2}\frac{\mathrm{d}\rho}{\mathrm{d}\varphi} = -r^2\frac{\mathrm{d}\rho}{\mathrm{d}\varphi},$$

we find from (11.11) that

$$\dot{r} = -\frac{L}{m}\frac{\mathrm{d}\rho}{\mathrm{d}\varphi}, \qquad \dot{r}^2 = \frac{L^2}{m^2}\left(\frac{\mathrm{d}\rho}{\mathrm{d}\varphi}\right)^2. \tag{11.12}$$

Substituting this expression for $\dot{r}^2$ into (11.10), we obtain the following differential equation for $\rho(\varphi)$:

$$\left(\frac{\mathrm{d}\rho}{\mathrm{d}\varphi}\right)^2 + \rho^2 - \frac{2Gm^2M}{L^2}\rho = \text{const.} \tag{11.13}$$

The variables $\rho$ and $\varphi$ in this first order differential equation can be separated and its solution can be found by standard methods. However, a further simplification is achieved if we differentiate this equation with respect to $\varphi$, thus replacing it with the following second order differential equation:

$$\frac{\mathrm{d}^2\rho}{\mathrm{d}\varphi^2} + \rho = C, \tag{11.14}$$

where we introduced the notation $C = Gm^2M/L^2$.

Such a differential equation with constant coefficients is often encountered in various problems. For instance, a similar equation describes the motion of a harmonic oscillator. The general solution to the equation is well known:

$$\rho(\varphi) = C + A\cos(\varphi - \varphi_0), \tag{11.15}$$

where $A$ and $\varphi_0$ are the arbitrary constants, whose values (for the problem of oscillations) depend on the initial conditions. Returning to the original function $r = 1/\rho$, we obtain:

$$r(\varphi) = \frac{p}{1 + e\cos(\varphi - \varphi_0)}, \tag{11.16}$$

where

$$p = \frac{1}{C} = \frac{L^2}{Gm^2M}, \qquad e = \frac{A}{C}. \tag{11.17}$$

Equation (11.16) describes the required trajectory of a body in the Newtonian (inverse-square) gravitational field (see section 10.4 equation (10.14)). It is well known from analytical geometry that this equation is the equation of a conic section (ellipse, parabola, or hyperbola), that is, of the curve formed by the intersection of a circular cone and a plane. In (11.16), $\varphi - \varphi_0$ is the angle between the radius vector and the axis of symmetry of the trajectory. (This axis is directed from the center of force toward the nearest point of the trajectory.) If we choose the polar axis along this axis of symmetry, the constant $\varphi_0$ in (11.16) is zero.

The quantity $p = L^2/(Gm^2M)$ in (11.16) has the dimension of length. It is called the *semilatus rectum* of the conic (or the *orbital parameter* or *focal parameter* if the conic is an orbit). The dimensionless quantity $e$ in (11.16) is called the *eccentricity* of the conic section. For $e = 0$, (11.16) gives $r = p$, that is, the distance $r$ does not depend on $\varphi$. Thus, for $e = 0$ the orbit is a circle whose radius equals $p$. Otherwise, the focal parameter $p$ equals the distance between the center and the orbit at $\varphi - \varphi_0 = \pm\pi/2$, when $\cos(\varphi - \varphi_0) = 0$. (This is the geometric sense of the focal parameter.) For $e < 1$, (11.16) corresponds to an ellipse (figure 10.1), for $e = 1$ to a parabola (to a degenerate ellipse) and for $e > 1$ to a hyperbola.

The parameters of the orbit $p$ and $e$ can be expressed in terms of the dynamical constants of the motion, namely the total energy $E$ and angular momentum $L$, and the physical parameters $M$ and $m$. According to (11.17), the focal parameter $p$ depends only on the angular momentum $L$: $p = L^2/(GMm^2)$. Next we obtain the values of the major axis $a$ (for closed orbits) and eccentricity $e$.

Let $r_P$ be the distance between the center of force and the closest point of the trajectory (perihelion, perigee, or, generally, pericenter) and $v_P$ be the velocity at this point. (The velocity vector $\mathbf{v}$ is orthogonal to the radius vector $\mathbf{r}$ at this point.) Then the total energy $E$ can be written as follows:

$$E = \frac{1}{2}mv_P^2 - \frac{GMm}{r_P} = \frac{L^2}{2mr_P^2} - \frac{GMm}{r_P}. \tag{11.18}$$

We have expressed the velocity at the perigee in terms of the angular momentum $L = mr_P v_P$. From (11.16), we see that the perigee distance $r_P$ for an orbit with given parameters $p$ and $e$ equals $p/(1 + e)$. Substituting $r_P = p/(1 + e)$ and $L^2 = pGMm^2$ from (11.17) into (11.18), we obtain

$$E = -(1 - e^2)\frac{GMm}{2p} = -\frac{GMm}{2a}. \tag{11.19}$$

In the latter formula we have used the relation $a = p/(1 - e^2)$ between the semimajor axis $a$ and parameters $p$ and $e$ of the ellipse (see (10.16)). For finite closed (elliptical) orbits $e < 1$ and hence $a$ is positive. We see that closed orbits correspond to negative values of the total energy: $E < 0$. It follows from (11.19) that the semimajor axis $a$ for such a closed elliptical orbit depends only on total energy $E$: $a = -GMm/(2E)$. The semimajor axis $a$ and, consequently, the period of revolution $T$ (see the following section) are uniquely determined by the total energy $E$ and are independent of the angular momentum. The period of revolution is the same for any orbit with a given energy (or given semimajor axis), independent of the eccentricity $e$ of the ellipse. For the special case of circular orbits we have $r = a$ and (11.19) coincides with the expression $E = -GMm/(2r)$ for the total energy of a satellite in a circular orbit, (10.11).

With a value of zero for the total energy, $E = 0$, (11.19) gives $e = 1$. The conic section, (11.16), with $e = 1$ is a parabola. Positive values of the total energy, $E > 0$, correspond to eccentricities $e > 1$, that is, to hyperbolic trajectories described by (11.16) with $e > 1$.

If we apply the relation $a = p/(1 - e^2)$ to a hyperbola ($e > 1$), it gives a negative value for $a$. In this case the magnitude of $a$ has the geometric meaning of the distance between the two branches of the hyperbola (measured along the principal axis that passes through the foci). With this in mind, we can also use (11.19) with $a < 0$ for open hyperbolic orbits corresponding to positive values of the total energy $E$.

In the general case, we can find from (11.19) the expressions for the semimajor axis $a$ and eccentricity $e$ of the orbit in terms of constant values of the total energy $E$ and angular momentum $L$. Substituting $p = L^2/(GMm^2)$ into (11.19), we obtain:

$$a = -\frac{GMm}{2E}, \qquad e = \sqrt{1 + \frac{2EL^2}{G^2M^2m^3}}. \tag{11.20}$$

## 11.3 Kepler's third law

In order to prove Kepler's third law for elliptical orbits, we can calculate the period of revolution $T$ by dividing the area $S = \pi ab$ of the ellipse by the sectorial velocity $\mathrm{d}S/\mathrm{d}t$. According to (11.7), the sectorial velocity is proportional to the magnitude of the angular momentum: $L = 2m\mathrm{d}S/\mathrm{d}t$. Since the semiminor axis $b$ of the ellipse equals $b\sqrt{1 - e^2}$ (see (10.18)) and $a = p/(1 - e^2)$, we obtain for the period of revolution:

$$T = \frac{2m\pi p^2}{L\left(1 - e^2\right)^{3/2}}.$$

Substituting $L = m\sqrt{GMp}$ here, we see that focal parameter $p$ and eccentricity $e$ enter into the expression for $T$ in the form $p(1 - e^2)$ which is equal to the semimajor axis $a$. Consequently, the period of revolution along an elliptical orbit depends only on its semimajor axis $a$:

$$T = \frac{2\pi a^{3/2}}{\sqrt{GM}}. \qquad (11.21)$$

The equation states that the square of the period of revolution is proportional to the cube of the semimajor axis of the orbit. This is the familiar form of Kepler's third law. For the special case of circular orbits, (11.21) for the period of revolution (with $a = r$) was obtained earlier (see (10.9)).

As an example of how Kepler's third law can be used in practice, let us consider the following problem.

**Problem**. A projectile is launched vertically from the Earth's surface with an initial velocity that equals the circular velocity $v_c = \sqrt{gR}$ for a very low orbit. How long does the flight of the missile last from start to finish (i.e. when it strikes the ground)?

**Solution**. It follows from the law of energy conservation that the height of the highest point reached by the missile over the surface for $v_0 = v_c$ equals the Earth's radius $R$. The trajectory of the missile is a half of the rectilinear segment joining the center of the Earth and the upper point of the flight. This segment can be regarded as the limiting case of a very narrow elliptical orbit with foci at the ends of the segment. To make the motion along this orbit theoretically possible, we can imagine all the Earth's mass to be concentrated at its center. The major axis of the degenerate ellipse is $2R$. The period of revolution $T_0$ along this degenerate elliptical orbit would be the same as along the circular orbit whose diameter equals the major axis of this degenerate ellipse: $T_0 = 2\pi R/v_c = 2\pi\sqrt{R/g}$. We are interested in the time interval during which one half of the degenerate orbit is completed, namely, the half that is farther from the center of force.

To find this time, we can make use of Kepler's second law. We consider an elliptical orbit whose semimajor and semiminor axes are $a$ and $b$, respectively (figure 11.2). When the end of the radius vector moves along this ellipse, the radius vector sweeps out equal areas in equal times. We want to find the time interval $T_1$ during which the end of the radius vector passes along the remote half of the ellipse from one end of the minor axis to the other, sweeping out the corresponding area $S_1$ (the unshaded portion of the figure) which consists of half of the ellipse and a

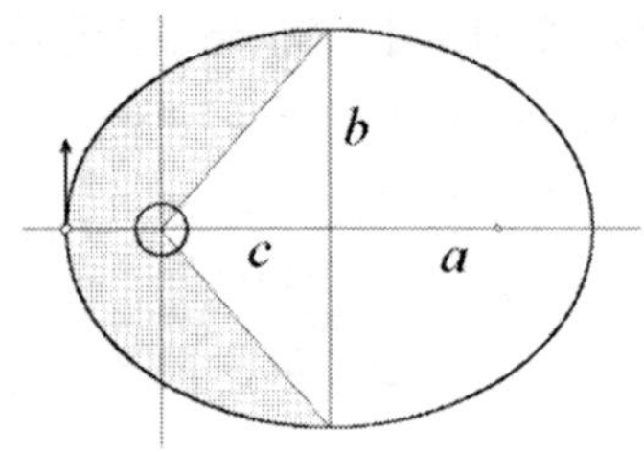


**Figure 11.2.** The area swept out by the radius vector.

triangle whose base is the minor axis and whose vertex is at the focus. The height $c$ of this triangle equals $ea$, where $e$ is the eccentricity of the ellipse.

Consequently, the ratio of the time interval $T_1$ to the period $T_0$ equals the ratio of corresponding areas:

$$\frac{T_1}{T_0} = \frac{S_1}{S_0} = \frac{\pi ab/2 + eab}{\pi ab} = \frac{1}{2} + \frac{e}{\pi}.$$

For the circular orbit $e = 0$ and we get, as expected, $T_1 = T_0/2$. That is, any half of the circular orbit is covered during one half of the period since the circular orbital motion is uniform. However, for elliptical orbits these halves of the ellipse are not equivalent: the farther half requires more time. For the limiting case of the degenerate ellipse (with $e = 1$), we obtain $T_1 = (1 + 2/\pi)T_0/2 = 0.82\ T_0$. During the first half of this time the missile moves upward and during the second half it falls. Since we are ignoring air resistance, the times of rising and falling are clearly equal.

## 11.4 A hodograph of the velocity vector for Keplerian motion

In chapter 3 we described the simulation program that illustrates an interesting property of any Keplerian motion. This property is concerned with the curve traced out by the velocity vector in velocity space (a hodograph). We have seen that for circular, elliptic and parabolic motion, the hodograph of the velocity is a circle (see figures 3.1 and 3.2) and for hyperbolic motion, it is a circular arc (figure 3.3). Here the circular shape of the velocity hodograph is proved rigorously on the basis of Newton's laws of motion.

This property of the velocity vector holds for the motion of a particle in the Newtonian central gravitational field (and the Coulomb central electrostatic field), whose strength decreases as the square of the distance. To prove the property analytically, we apply Newton's second law to the motion of a particle under the central force $F(r) = GmM/r^2$. The vector $\mathbf{a}$ of the acceleration produced by this force is always directed toward the center of force and its magnitude is inversely proportional to the square of the distance $r$:

$$\mathbf{a} = \frac{\Delta \mathbf{v}}{\Delta t} = -GM\frac{1}{r^2}\frac{\mathbf{r}}{r}. \tag{11.22}$$

We can eliminate the variable $1/r^2$ from this equation by using the law of angular momentum conservation (see section 11.1 'Angular momentum and areal velocity'):

$$L = m|\,\mathbf{r} \times \mathbf{v}\,| = mrv_{\perp} = mr^2\dot{\varphi}. \tag{11.23}$$

Here $v_{\perp} = r\dot{\varphi}$ is the transverse component of the velocity of the particle (the component orthogonal to the radius vector). Since the angular momentum of the particle remains constant during the motion in a central force field, for any point of the trajectory the square of the distance $r$ from the origin can be expressed with the help of (11.23) through the angular velocity $\dot{\varphi} = \Delta\varphi/\Delta t$ and a constant value of $L$:

$$\frac{1}{r^2} = \frac{m}{L}\dot{\varphi} = \frac{m}{L}\frac{\Delta\varphi}{\Delta t}. \tag{11.24}$$

Substituting $1/r^2$ given by (11.24) into (11.22), we find that during Keplerian motion the magnitude of the infinitesimal vector $\Delta\mathbf{v}$ (of the increment in the velocity vector during the time interval $\Delta t$) is proportional to the angle $\Delta\varphi$, through which the radius vector of the particle rotates during $\Delta t$:

$$|\Delta\mathbf{v}| = \frac{GMm}{L}\Delta\varphi. \tag{11.25}$$

We note that this proportionality between $|\Delta\mathbf{v}|$ and $\Delta\varphi$ holds only for the motion in a central field whose strength is proportional to $1/r^2$. Hence the circular form of the velocity hodograph (the property that we are going to prove) is an inherent property of inverse-square central fields.

A geometric interpretation of (11.25) is shown in figure 11.3. Each time radius vector $\mathbf{r}$ of the orbiting particle turns through an infinitesimal angle $\Delta\varphi$, the vector of velocity $\mathbf{v}$ is incremented by $\Delta\mathbf{v}$, whose magnitude $|\Delta\mathbf{v}|$ is proportional to $\Delta\varphi$. Thus the elementary vectors $\Delta\mathbf{v}$ in velocity space lie along a circle whose radius, according to (11.25), equals $GMm/L$. In other words, (11.25) proves that the hodograph of the velocity vector is a circle. The constant value $u = GMm/L$ of the coefficient of proportionality between $|\Delta\mathbf{v}|$ and $\Delta\varphi$ is the radius of this circular hodograph of the velocity vector.

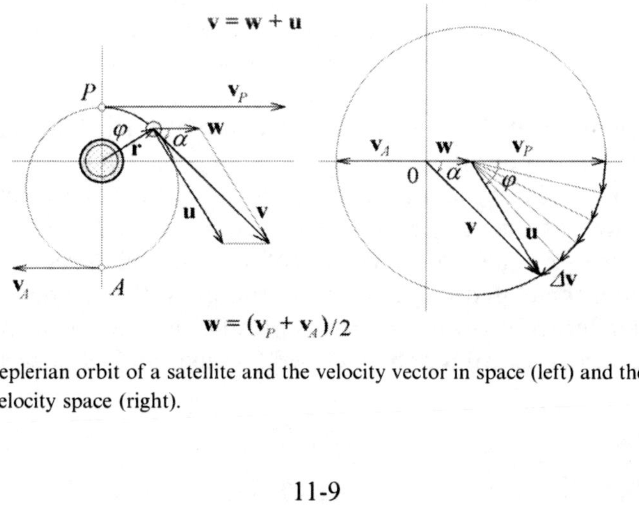


**Figure 11.3.** The Keplerian orbit of a satellite and the velocity vector in space (left) and the hodograph of the velocity vector in velocity space (right).

It is convenient to express radius $u$ of the velocity hodograph in terms of velocity $v_P$ at the perigee (the point $P$ nearest to the center of force) and the circular velocity $v_c$ for the perigee distance $r_P$. Since $v_c = \sqrt{GMm/r_P}$ and $L = mr_P v_P$, we obtain the expression:

$$u = \frac{v_c^2}{v_P}. \tag{11.26}$$

If the satellite is launched into a circular orbit ($v_P = v_c$), (11.26) yields $u = v_c$. In this trivial case the radius of the hodograph clearly equals the circular velocity. For an elliptical orbit, it is possible to express the radius $u = GMm/L$ of the hodograph in terms of velocities $v_P$ and $v_A$ at the perigee and apogee, respectively. We can use the laws of conservation of energy and angular momentum for this purpose. Equating the values of the total energy at these points $r_P$ and $r_A$, we write:

$$\frac{m}{2}v_P^2 - \frac{GMm}{r_P} = \frac{m}{2}v_A^2 - \frac{GMm}{r_A}. \tag{11.27}$$

Next, we solve this equation for $GMm$ and substitute into $u = GMm/L$ the expression obtained, together with the (constant) angular momentum $L$, calculated, say, for the perigee: $L = mr_P v_P$. The ratio of distances $r_P/r_A$ can be eliminated with the help of the relation $r_P v_P = r_A v_A$. (The values of the angular momentum, or of the sectorial velocity, are equal at the perigee and apogee). Finally, we obtain the following expression for the radius $u$ of the velocity hodograph:

$$u = \frac{1}{2}(v_P + v_A). \tag{11.28}$$

By virtue of this property we can represent the vector of velocity $\mathbf{v}$ for any point of an elliptical orbit as the vector sum of the following two vectors $\mathbf{w}$ and $\mathbf{u}$. (See figure 11.3 for the case of an elliptical orbit.) One term of the sum is the constant vector $\mathbf{w} = (\mathbf{v}_P + \mathbf{v}_A)/2$ of magnitude $(v_P - v_A)/2$, directed along the vector $\mathbf{v}_P$ of the velocity at perigee. This vector $\mathbf{w}$ extends from the origin of velocity space to the center of the circular hodograph. The second term is a vector $\mathbf{u}$ of constant magnitude $u = (v_P + v_A)/2$, whose direction is always perpendicular to the radius vector $\mathbf{r}$ of the orbiting body.

This representation of the velocity vector is useful for solving certain problems concerning Keplerian motion.

The derivation of (11.25) that gives the proof of the circular form of the velocity hodograph, as well as the geometric interpretation of (11.25) discussed above, are based on the inverse-square dependence of the central force on distance. The assumption about the closed trajectory is used only in the calculation of the hodograph radius $u$ in terms of $v_P$ and $v_A$. Consequently, the circular form of the velocity hodograph is characteristic not only of closed (circular and elliptical) orbits, but also

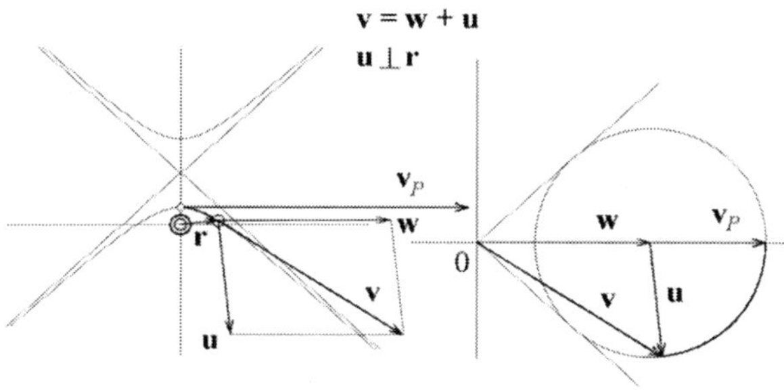


**Figure 11.4.** The hyperbolic trajectory of a body in a central gravitational field and the velocity vector in space (left), and the hodograph of the velocity vector in velocity space (right).

of open (parabolic and hyperbolic) trajectories of motion in a central field whose force is inversely proportional to the square of the distance from the center.

Motion along the parabolic trajectory corresponds to the limit $v_A \to 0$ in the above formulas. In this case $\mathbf{w} = \mathbf{v}_P/2$ and both terms of the sum $\mathbf{u} + \mathbf{w}$ have equal magnitudes: $u = w = v_P/2$. These vectors **w** and **u** form two radii of the circular hodograph and at any instant the vector of velocity **v** forms a chord of the circle.

For a hyperbolic Keplerian motion, the diameter $2u$ of the circular hodograph is smaller than the velocity $v_P$ at the vertex of the hyperbola (figure 11.4). In this case the origin of velocity space is located outside the circular hodograph.

When a body moves from infinity toward the center of force and then again to infinity along a hyperbolic trajectory, the vector of velocity rotates through the angle between the asymptotes of the hyperbola. In velocity space, the end of the velocity vector traces the part of the circle bounded by the points of tangency with this circle of the straight lines drawn from the origin of the velocity space. (See figure 11.4.) The end of vector **u** of constant magnitude $v_c^2/v_P$ moves along the same arc. This vector rotates (non-uniformly) between the points of tangency about a fixed point (the center of the hodograph) whose position with respect to the origin of the velocity space is indicated by the constant vector **w**.

## 11.5 Another derivation of Kepler's first law

Traditional derivations of the equation for a planetary orbit in university mechanics courses are usually based on using the conservation laws of angular momentum and the total energy, or on a transformation of the differential equation of motion by introducing another unknown function $1/r$ instead of $r(\theta)$ (see, for example, section 11.2 above). For most undergraduate students, the first method requires a severe struggle through the mathematics, while the second may seem rather artificial.

However, there exists another very succinct and elegant way to obtain the polar equation of the orbit, $r = p/(1 + e\cos\theta)$. This method is based on using the velocity space. The derivation of the Keplerian orbit is performed below together with another proof of the velocity hodograph circularity for any orbit in an inverse-square central field.

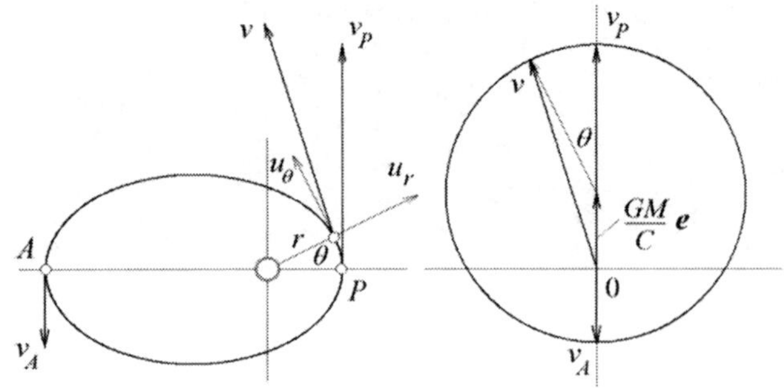


**Figure 11.5.** The Keplerian orbit and the velocity vectors in space (left), and the hodograph of the velocity vector in velocity space (right).

The derivation is quite straightforward. It starts with Newton's second law of motion in the form which states that under the central force of gravity the velocity $\mathbf{v} = \mathrm{d}\mathbf{r}/\mathrm{d}t$ of the celestial body varies in accordance with the following equation:

$$\frac{\mathrm{d}\mathbf{v}}{\mathrm{d}t} = -\frac{GM}{r^2}\mathbf{u}_r, \tag{11.29}$$

where $\mathbf{u}_r = \mathbf{r}/r$ is the unit radial vector (see the left panel of figure 11.5). We also introduce another unit vector $\mathbf{u}_\theta$ orthogonal to $\mathbf{u}_r$.

Then $r^2$ is eliminated from (11.29) with the help of the angular momentum conservation $L = mr^2\dot{\theta}$, which yields the following equation:

$$\frac{\mathrm{d}\mathbf{v}}{\mathrm{d}t} = -\frac{GM}{C}\dot{\theta}\mathbf{u}_r. \tag{11.30}$$

Here a constant $C = L/m = r^2\dot{\theta}$ is introduced whose meaning is the angular momentum magnitude per unit mass, or the sectorial velocity doubled ($C = 2\mathrm{d}S/\mathrm{d}t$).

The crucial point of the derivation is the substitution of $\mathrm{d}\mathbf{u}_\theta/\mathrm{d}t$ in (11.30) instead of $-\dot{\theta}\mathbf{u}_r$ (see figure 11.5), which yields

$$\frac{\mathrm{d}\mathbf{v}}{\mathrm{d}t} = \frac{GM}{C}\mathrm{d}\mathbf{u}_\theta/\mathrm{d}t. \tag{11.31}$$

Subsequent straightforward integration of (11.31) with respect to time $t$ gives the expression for vector $\mathbf{v}$ as the sum of two vectors:

$$\mathbf{v} = \frac{GM}{C}(\mathbf{u}_\theta + \mathbf{e}). \tag{11.32}$$

The second term in the right-hand side of equation (11.32)—the constant of integration—is a time independent vector of magnitude $(GM/C)e$, while the first one is a vector of constant magnitude ($GM/C$), currently pointing in the direction of the unit vector $\mathbf{u}_\theta$, which is perpendicular to the momentary radius vector $\mathbf{r}$.

Equation (11.32) actually proves that the velocity hodograph for an arbitrary Keplerian motion is a circle (see the right panel of figure 11.5). Indeed, the unit vector $\mathbf{u}_\theta$ changes its direction as the body moves along its orbit and hence the vector $(GM/C)\mathbf{u}_\theta$ of fixed magnitude $GM/C$ rotates (non-uniformly) in the velocity space about the point at which the constant vector $(GM/C)\mathbf{e}$ points. Since $\mathbf{v}$ is the sum of these two vectors, its end generates the same circle (or an arc of the circle for hyperbolic orbits). This statement is equally valid for all closed (elliptical) and open (parabolic and hyperbolic) orbits traced under the inverse-square central force.

A similar, although less straightforward, derivation of the circular shape of the velocity hodograph (based primarily on geometrical considerations) was presented above (see section 11.4). The velocity vector of a body in an arbitrary Keplerian motion is also represented there as the sum of two vectors ($\mathbf{v} = \mathbf{u} + \mathbf{w}$ in notations used in section 11.4), one of which ($\mathbf{w} = (GM/C)\mathbf{e}$) always points from the origin to the same point of velocity space (the center of the hodograph), while the other vector of a constant magnitude ($\mathbf{u} = (GM/C)\mathbf{u}_\theta$) generates the circle.

When the constant component $\mathbf{w}$ of the velocity vector (which points to the center of the hodograph) is denoted by $(GM/C)\mathbf{e}$, the magnitude $e$ of vector $\mathbf{e}$ has a clear physical meaning of the eccentricity of the orbit. This makes it reasonable to call $\mathbf{e}$ the *eccentricity vector*.

The constant magnitude $GM/C$ of the other vector $\mathbf{u} = (GM/C)\mathbf{u}_\theta$ (radius of the circular hodograph) can be conveniently expressed in terms of velocity $v_P$ at the perihelion (perigee) and the circular velocity $v_c = \sqrt{GM/r_P}$ for this point $P$ of the orbit (see section 11.4): $u = GM/C = v_c^2/v_P$. The displacement $w$ of the hodograph center from the origin of the velocity space can be expressed as $v_P - v_c^2/v_P$ or, equivalently, as $w = ue$, where the eccentricity $e = v_P^2/v_c^2 - 1$.

The last step, which allows us to obtain the orbit from (11.32), is rather obvious: it consists of taking a projection of both sides of (11.32) in the direction of the unit vector $\mathbf{u}_\theta$. From the left panel of figure 11.5 we see that this projection of $\mathbf{v}$ equals $r\dot{\theta}$ or $C/r$ (if we take into account that $r^2\dot{\theta} = C$). The right panel of the figure shows that at the same time this projection equals $(GM/C)(1 + e\cos\theta)$. Equating these values, we obtain the desired equation of the orbit:

$$r = \frac{p}{1 + e\cos\theta}, \quad \text{where} \quad p = \frac{C^2}{GM} = \frac{L^2}{GMm^2} = r_p(1 + e). \tag{11.33}$$

The geometrical method from equation (11.32) to (11.33) described in this section may seem to undergraduate students a more natural dynamical derivation of Kepler's first law compared to the highly mathematical derivation in section 11.2.

## 11.6 A family of orbits with equal energies and a common initial point

Earlier in section 4.1.2, we discussed the properties of the set of orbits traced by the fragments of an exploding rocket. We assumed that all these fragments initially move from a single point in all directions with initial velocities of equal magnitude. They are thus satellites of the Earth (or ballistic projectiles), orbiting along different elliptical Keplerian orbits characterized by equal values of total energy (per unit mass). These orbits are limited in space to a region whose boundary is a closed surface of revolution. The axis of rotation of the curve generating this surface passes through the Earth's center and the initial point (see figure 4.5). Next we find the equation that defines this curve and show how it can be used to solve some practical problems.

### 11.6.1 The envelope surface for the family of orbits

We show below that the surface is an ellipsoid, generated by the rotation of an ellipse about the axis mentioned above. One of the foci of the ellipse is located at the center of the Earth and the other at the initial point. The dimensions and eccentricity of the ellipsoid depend on the position of the initial point and on the initial velocity of the fragments.

The fragment whose initial velocity, at point $S$ in figure 11.6, is directed upward along the local vertical line (to the left in figure 11.6), rises vertically along a straight line to the highest point $N$ located at the distance $r_{\max}$ from the Earth's center. Then it falls toward the Earth along the same line. The trajectory of this fragment is part of the rectilinear segment joining the highest point $N$ with the center of the Earth $F_1$. We can consider this segment as the limiting case of an infinitely narrow ellipse. The foci of this degenerate ellipse lie at the ends of the segment. That is, one focus is at the Earth's center $F_1$ and the other at the highest point $N$ of the trajectory.

Clearly the highest point $N$ is on the bounding surface. The distance $r_{\max}$ of this point from the Earth's center can be calculated by equating the total energy of the fragment at this point $N$ to the total energy at the initial point $S$, located at the distance $r_0$ from the center of force:

$$\frac{v_0^2}{2} - \frac{GM}{r_0} = -\frac{GM}{r_{\max}}. \qquad (11.34)$$

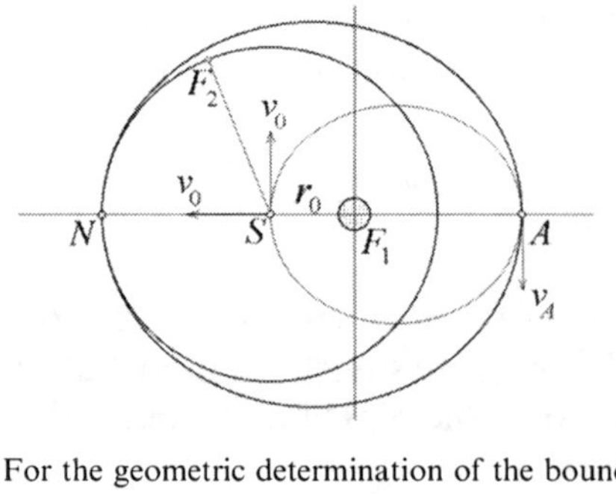


**Figure 11.6.** For the geometric determination of the boundary surface.

Here, it is convenient to express the gravitational parameter of the planet $GM$ in terms of the escape velocity $v_{esc}$ for the initial point $S$ ($v_{esc}^2 = 2GM/r_0$):

$$\frac{1}{r_{max}} = \frac{1}{r_0}\left(1 - \frac{v_0^2}{v_{esc}^2}\right), \qquad r_N = r_{max} = \frac{r_0}{1 - \left(v_0/v_{esc}\right)^2}. \tag{11.35}$$

If the initial velocity equals the circular velocity for the initial point, that is, if $v_0 = v_c = \sqrt{GM/r_0}$, (11.35) gives $r_{max} = 2r_0$: the distance of the highest point $N$ from the Earth's center is twice the distance $r_0$ of the initial point.

We can easily find a second point on the boundary, namely, the point opposite $N$. It coincides with the apogee $A$ (or with the perigee if $v_0 < v_c$) of the elliptical orbit of that fragment whose initial velocity at $S$ is directed horizontally (transverse to the radius vector). The distance $r_A$ of this point from the Earth's center is given by (10.22) of the preceding chapter:

$$r_A = \frac{r_0}{2\left(v_c/v_0\right)^2 - 1}. \tag{11.36}$$

Next we prove that the curve whose rotation generates the boundary is an ellipse. The ends of the major axis of this ellipse are located at $N$ and $A$, and its foci are located at the initial point $S$ and the Earth's center $F_1$. The shape of the boundary is determined from the following geometric properties of the trajectories.

First we find the locus of the foci of all the orbits of the fragments. All of the orbits have a focus at the center of the Earth and so the locus of this set is the point $F_1$. The locus of the set of second foci $F_2$ is a circle whose center is located at the initial point $S$, and whose radius is equal to the distance $SN$, measured from $S$ to the most remote point $N$ (see figure 11.6). Indeed, for any orbit of the family, the sum of two distances of each point on the orbit from the foci of the orbit equals the major axis of the orbit. The major axes of all the orbits of the family, as we have seen, are equal to each other. Their lengths are equal to the length $r_{max}$ of the segment $F_1N$. This segment can be considered the major axis of the degenerate elliptical orbit of the fragment whose initial velocity is directed upward. All orbits of the family pass through $S$ and the distance from this common point to the focus $F_1$ for all the orbits equals $r_0$. Consequently, the distance between $S$ and the second focus also must be equal for all the orbits. Hence the second foci of all of the orbits of the family lie on the circle with the center at the initial point $S$ and radius $SN$.

Next we consider the following auxiliary construction (figure 11.7): we draw a second circle with the center at the Earth's center $F_1$ and radius $r_{max}$. This circle passes through $N$, which lies on the bounding surface.

Now let us consider the problem of finding the orbit passing through an arbitrary point $M$ that lies within the second circle just drawn. Choosing $M$ as a center, we draw a third circle tangent to the second circle at point $B$, as shown in figure 11.7. The second focus of the orbit through $M$ must lie on this circle because the sum of two distances from the foci again must be equal to $r_{max}$. And at the same time the

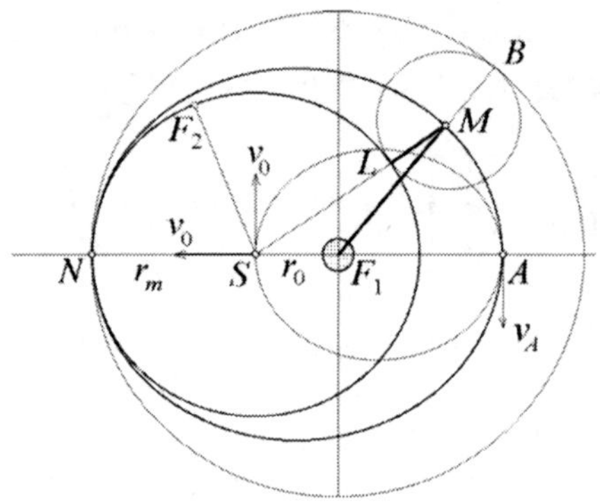


**Figure 11.7.** Determining the bounding surface.

second focus must lie on the first circle (with the center at the initial point $S$ and radius $SN$). We examine three possibilities:

1. The third circle (with the center at $M$) *intersects* the first circle (the locus of the second foci of the orbits) at two points. Then there exist *two orbits* of the family that pass through the given point $M$. The second foci of these two orbits lie at the two points of intersection.
2. The third circle has *no common points* with the first circle. Then no orbit of the family passes through $M$. If follows that $M$ lies *outside* the bounding surface.
3. Lastly, the third circle *grazes* the first circle, thus having a single common point $L$ with it (see figure 11.7). Then *only one orbit* of the family passes through $M$. In this case point $M$ must lie on the bounding surface. We can see from figure 11.7 that in this case of grazing, the sum of distances from $M$ to $F_1$ and $S$ equals the radius $r_{\max}$ of the second circle plus the radius $SN$ of the first circle. This sum is independent of the position of point $M$ on the boundary. That is, the sum has equal values for all points of the boundary. Since points $F_1$ and $S$ are fixed, and since the sum of their distances from $M$ is the same for all $M$, we have proved that the locus of the boundary points for the region occupied by the orbits of the family is an ellipse whose foci are at $F_1$ and $S$.

The eccentricity of the bounding ellipse can be found as the ratio of the distance $r_0$ between its foci to the major axis $r_N + r_A$. Using equations (11.35) and (11.36), we find:

$$e = \frac{r_0}{r_N + r_A} = \frac{v_{\mathrm{esc}}^2 - v_0^2}{v_{\mathrm{esc}}^2 + v_0^2}. \tag{11.37}$$

If we increase the initial speed of the fragments, the bounding surface expands and its shape becomes almost spherical. Indeed, as we can see from equation (11.37), its eccentricity $e$ becomes smaller and tends to zero as the initial speed approaches the escape velocity. If the initial velocity $v_0$ equals the circular velocity $v_c$ for the initial point, the distance between the foci of the bounding ellipse is one third its major axis. That is, the eccentricity of the ellipse is 1/3 if $v_0 = v_c$. If the initial speed tends to zero, the eccentricity of the envelope approaches unity. The apexes (the ends of the major axis) of the bounding ellipse approach its foci, the ellipse becomes very narrow and stretched through being spanned over the initial point $S$ and the center of the Earth $F_1$. This limiting case of a degenerate ellipse corresponds to the well known parabolic shape of the bounding surface ('safety paraboloid') for the trajectories of fragments moving within a restricted spatial region (in the vicinity of the initial point $S$) in which the gravitational field can be regarded as uniform.

We note that the exploitation of the geometrical properties of ellipses allowed us to easily find the envelope for the given family of orbits without tedious calculations.

### 11.6.2 Applications of the envelope surface

Knowing the boundary surface can be very useful in solving various problems concerning orbital motion, space dynamics and ballistics. For example, we can easily find the minimal firing speed of a projectile needed to hit a given target from a given starting point. Suppose we have a target $M$ (see figure 11.7) at a given location, which is determined by distance $r_M = |F_1M|$ from the force center $F_1$ (the center of the Earth) and distance $l_M = |SM|$ from the given starting point $S$ (whose distance from the force center is $r_0 = |F_1S|$). What is the minimal initial speed and what should the firing angle be?

Obviously, the firing speed is minimal if the target $M$ lies on the bounding surface. Since this boundary is an ellipse, the sum of distances $|F_1M|$ and $|SM|$ from $M$ to its foci (points $F_1$ and $S$) is equal to the major axis of the bounding ellipse: $|F_1M| + |SM| = r_N + r_A$. The sum $|F_1M| + |SM|$ is just the sum of given distances $r_M$ and $l_M$ to the target from $F_1$ and $S$. Let us denote this sum as $b$: $r_M + l_M = b$. Thus, we can equate this given value $b$ to the major axis $r_N + r_A$, which has already been calculated above when we derived equation (11.37) for the eccentricity of the bounding ellipse:

$$b = r_N + r_A = r_0 \frac{v_{\text{esc}}^2 + v_0^2}{v_{\text{esc}}^2 - v_0^2}. \tag{11.38}$$

Solving equation (11.38) for $v_0$, we obtain the desired minimal firing speed:

$$v_{0\,\text{min}}^2 = v_{\text{esc}}^2 \frac{b - r_0}{b + r_0}. \tag{11.39}$$

Equation (11.39) shows that for a given position of the starting point $S$ the minimal firing speed depends only on $b$, that is, on the sum of distances $r_M$ and $l_M$ that determine the target location ($b = r_M + l_M$). According to equation (11.39), the firing speed is zero if $b = r_0$, that is, for any target that lies on the segment $SF_1$ joining the starting point and the center of the Earth. The minimal firing speed $v_0$ tends to the escape velocity $v_{\text{esc}} = \sqrt{2gR^2/r}$ as the target is moved away to infinity (as $b \to \infty$).

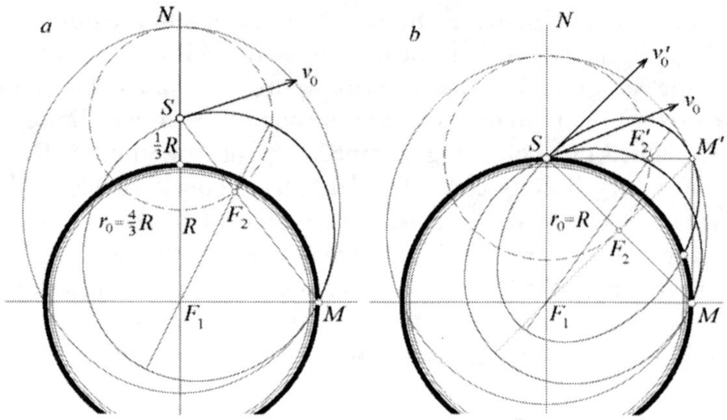


**Figure 11.8.** Examples of trajectories that correspond to the minimal starting speed of the projectile for a given starting point $S$ and a given target $M$.

The trajectory of the projectile fired to the given target $M$ with the minimal starting speed is a portion of the ellipse passing through $S$ and $M$. One focus of this Keplerian ellipse is located at the center of the Earth, while the second focus belongs to the circle whose center is at the starting point $S$ and whose radius equals $|SN|$ (see figure 11.7). Therefore this second focus is located at point $L$ at which the segment $SM$ from the starting point to the target intersects the circle mentioned. Knowing the locations of both foci for the elliptical trajectory of the projectile, we can easily find the firing angle with the help of the optical property of the ellipse. Since the light ray emitted from the focus $F_1$ to $S$ must be reflected at $S$ by the elliptical mirror toward the second focus $L$, the tangent to the ellipse at $S$ (and hence the direction of initial velocity $\mathbf{v}_{0\,\min}$) is the bisectrix of the angle $LSN$ (or $MSN$, see figure 11.7).

Particular examples of trajectories traced by the projectiles launched with minimal initial speeds to given targets are shown in figure 11.8. Let the starting point $S$ be located at the height of one third of the Earth's radius $R$ over the North Pole and the target be located on the equator (figure 11.8($a$)). Therefore $r_0 = \frac{4}{3}R$, $r_M = R$ and $l_M = \frac{5}{3}R$, so that $b = \frac{8}{3}R$. For the minimal starting speed in this case (11.39) yields $v_{0\,\min}^2 = \frac{2}{3}v_{\text{circ}}^2$, $v_{\min} = 0.8165\ v_{\text{circ}}$. From the triangle $F_1SM$ we can see that the sine of angle $MSN$ equals $\frac{3}{5}$. The initial velocity $\mathbf{v}_0$ must be directed along the bisectrix of this angle. Hence the angle between vector $\mathbf{v}_0$ and the upward vertical line must equal 71.565°. The trajectory of this projectile is a portion $SM$ of an ellipse whose foci are located at $F_1$ (the Earth's center) and $F_2$. The latter point lies on the straight line $SM$ joining the starting point and the target.

We note that at the target point $M$ (see figure 11.8) both ellipses (the trajectory and the envelope bounding surface) have a common tangent. According to the optical property, the ray $F_1M$ emitted from the common focus $F_1$ of these ellipses must be reflected at $M$ by both curves toward their second foci ($F_2$ and $S$, respectively). Therefore all three points ($M$, $F_2$ and $S$) lie on the same straight line $MS$.

If the starting point $S$ is located on the surface of the Earth (figure 11.8($b$)), for the same target $M$ on the equator we have $l_M = \sqrt{2}R$, $b = (1 + \sqrt{2})R$ and (11.39) yields $v_{0\,\min} = 0.9102\ v_{\rm circ}$. The angle $MSN$ in this case equals 135°, so that vector $\mathbf{v}_0$ must make an angle of 67.5° with the upward vertical.

The same minimal starting speed is required for the target $M'$ located in space at the distance $R$ on the straight line directed horizontally from point $S$ (see figure 11.8($b$)). We can make this conclusion either analytically from equation (11.39), or geometrically from the observation that both targets $M$ and $M'$ belong to the same bounding surface. To hit this target, the vector $\mathbf{v}_0$ must make an angle of 45° with the upward vertical. We note that on the surface of the Earth, the range of this projectile is smaller than in the preceding case in which the projectile is fired with the same starting speed at an angle of 67.5°. Compare this result with the commonly known fact that in approximations of the 'flat Earth' (a uniform field of gravity), a starting angle of 45° gives the greatest possible range for the projectile.

In this section we have tried to show that interesting and useful peculiarities of the families of Keplerian orbits can be obtained using very modest means based solely on the fact that these orbits are ellipses characterized by commonly known simple geometrical properties. Many demanding problems concerning orbital motion can easily be solved with the help of these properties. Simple geometrical considerations can provide elegant solutions which allow us to avoid complicated and tedious calculations.

## 11.7 Relative orbital motion

To investigate the relative motion of orbiting bodies we can derive simple differential equations that are approximately valid for small spatial distances between the bodies. More precisely, these equations describe the relative motion of the orbiting bodies while the distances between them are much smaller than the linear dimensions (axes) of the orbit.

As an example, we consider the motion of a body ejected by an astronaut from an orbital station that orbits the Earth in a circle. (Several simulation experiments of this kind are described in section 5.3.) We make use of the non-inertial frame of reference whose origin lies in the station (figure 11.9). The $z$-axis of this frame points perpendicularly to the plane of the orbit, that is, parallel to the vector of the angular velocity $\Omega$ of revolution of the station; the $x$-axis lies in the plane of the orbit and extends radially outward, away from the center of the Earth; and the $y$-axis is parallel to the orbital velocity, $\mathbf{v}_c$. Vector $\mathbf{r}_0$ is directed from the center of the Earth toward the orbital station. The position of the body relative to the orbital station is determined by the radius vector $\mathbf{r}$ and relative to the center of the Earth by a vector $\mathbf{r}'$, that is the sum of $\mathbf{r}_0$ and $\mathbf{r}$: $\mathbf{r}' = \mathbf{r}_0 + \mathbf{r}$.

The acceleration $\mathbf{a}$ of the body relative to this rotating frame of reference is determined by the gravitational pull of the Earth $-GmM\ \mathbf{r}'/r'^3$ (here $M$ is the mass of the Earth and $m$ is the mass of the body), and also by the centrifugal pseudo-force of inertia $-m\boldsymbol{\Omega} \times (\boldsymbol{\Omega} \times \mathbf{r}')$ and by the Coriolis pseudo-force of inertia

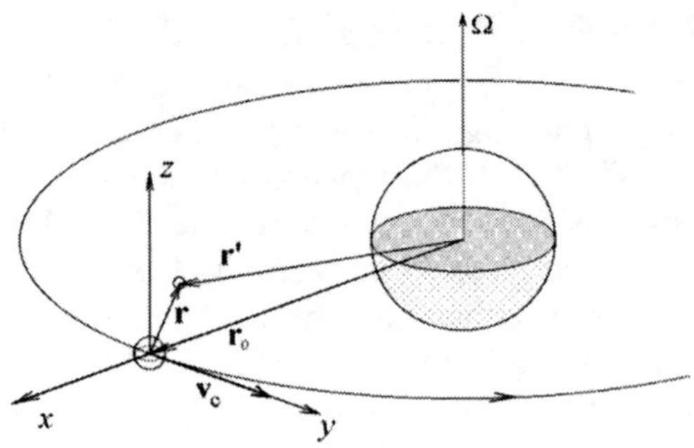


**Figure 11.9.** The frame of reference associated with the orbital station.

$2\,m\mathbf{v} \times \mathbf{\Omega}$, where $\mathbf{v}$ is the vector of velocity of the body relative to the station. Thus,

$$\mathbf{a} = -\frac{GM}{r'^3}\mathbf{r}' - \mathbf{\Omega} \times (\mathbf{\Omega} \times \mathbf{r}') + 2\mathbf{v} \times \mathbf{\Omega}. \tag{11.40}$$

Now we represent vector $\mathbf{r}'$ in (11.40) as the sum $\mathbf{r}_0 + \mathbf{r}$ (see figure 11.9). To simplify the expression $1/r'^3$ in the case of small spatial distances $r \ll r_0$, we first calculate the square of the vector $\mathbf{r}'$ and then raise the expression obtained to the power $-3/2$. Calculating the square of the vector $\mathbf{r}_0 + \mathbf{r}$, we reject the small term $r^2$. Leaving only the terms linear in small quantity $r/r_0$, for the first term in the right side of (11.40) we obtain:

$$\frac{GM}{r'^3} \approx \frac{GM}{r_0^3}\left(1 + 2\,\frac{\mathbf{r}_0 \cdot \mathbf{r}}{r_0^2}\right)^{-3/2} \approx \Omega^2\left(1 - 3\,\frac{\mathbf{r}_0 \cdot \mathbf{r}}{r_0^2}\right). \tag{11.41}$$

We have taken into account here that the square of the angular velocity $\Omega$ of the station in its orbital motion equals $GM/r_0^3$. Substituting (11.41) into (11.40), we obtain the following expression for the relative acceleration of the body, valid up to the terms linear in $r/r_0$:

$$\mathbf{a} = -\Omega^2(\mathbf{r}_0 + \mathbf{r}) + 3\Omega^2\frac{\mathbf{r}_0 \cdot \mathbf{r}}{r_0^2}\,\mathbf{r}_0 - \mathbf{\Omega} \times (\mathbf{\Omega} \times \mathbf{r}_0) - \mathbf{\Omega} \times (\mathbf{\Omega} \times \mathbf{r}) + 2\mathbf{v} \times \mathbf{\Omega}. \tag{11.42}$$

In (11.42), the main term of the acceleration caused by the gravitational pull of the Earth ($-\Omega^2\mathbf{r}_0$) is balanced by the main term of the acceleration caused by the centrifugal pseudo-force of inertia, namely by the term $-\mathbf{\Omega} \times (\mathbf{\Omega} \times \mathbf{r}_0)$. This relationship is clearly seen from figure 11.9. The balancing of the gravitational force by the pseudo-force of inertia is the sense of weightlessness experienced by astronauts on the orbital station. The balancing is complete for a body located at the origin of the non-inertial frame of reference associated with the station.

For a body at some distance $\mathbf{r}$ from the origin, only the main term of the gravitational acceleration (of order zero in $r/r_0$) is balanced by the pseudo-force of inertia.

The remaining terms of gravitational acceleration (linear in $r/r_0$) in (11.42), together with the terms of the same order of magnitude in the acceleration caused by the centrifugal pseudo-force of inertia, and with the acceleration produced by Coriolis force (which depends on the relative velocity $\mathbf{v}$), give the desired differential equations of motion relative to the orbital station.

Projections of the vectors in (11.42) onto the axes of the reference frame associated with the station give the following system of differential equations describing approximately the relative motion:

$$\begin{aligned} \ddot{x} &= 3\Omega^2 x + 2\Omega\, \dot{y}, \\ \ddot{y} &= 2\Omega\, \dot{x}, \\ \ddot{z} &= -\Omega^2 z. \end{aligned} \tag{11.43}$$

Here $x$, $y$ and $z$ are the components of radius vector $\mathbf{r}$ which determines the position of the body relative to the station, and $\dot{x}$, $\dot{y}$ and $\dot{z}$ are the components of the relative velocity.

The motion of the body ejected from the station starts from the origin of the non-inertial reference frame. Therefore, in all cases under consideration $x(0) = 0$, $y(0) = 0$ and $z(0) = 0$. Next we find particular solutions of (11.43) for different directions of the initial velocity $\Delta\mathbf{v}$ of the body relative to the station.

1. The body is ejected from the station in a direction perpendicular to the plane of the orbit; that is, $\dot{x}(0) = 0$, $\dot{y}(0) = 0$ and $\dot{z}(0) = \Delta v$. For these initial conditions, the particular solution of (11.43) describes rectilinear oscillatory motion along the $z$-axis: $x(t) = 0$, $y(t) = 0$ and $z(t) = (\Delta v/\Omega)\sin \Omega t$. That is, the body moves sinusoidally relative to the station along the $z$-axis, with the station at the center of the motion and with a period $T = 2\pi/\Omega$. This period is equal to the period of revolution of the station along its circular orbit. At every quarter of this revolution the body is at its maximum distance from the station: $l = \Delta v/\Omega = r_0(\Delta v/v_c)$.

As an example we let the height of the circular orbit of the station be $h = 0.1\ R \approx 640$ km (the radius of the orbit $r_0 \approx 7000$ km and the period of revolution $T \approx 98$ min) and the relative initial velocity of the body be $\Delta v = 15$ m s$^{-1}$, which is approximately 0.2% of the orbital velocity $v_c = 7.5$ km s$^{-1}$. In this case the maximal distance $l$ of the body from the station is approximately 14 km.

This result can be easily understood in terms of the motion of the body in the geocentric frame of reference. In this case the body moves along an almost circular orbit that differs from the orbit of the station only by a slightly different orientation of its plane. The angle between these planes equals the (small) angle $\alpha$ between vector $\mathbf{v}_c$ of the circular velocity of the station and vector $\mathbf{v}_c + \Delta\mathbf{v}$ of the geocentric initial velocity of the body. Here $\alpha \approx \Delta v/v_c \approx 2 \cdot 10^{-3}$. Making revolutions along their orbits with practically equal periods, the body and the station meet after every half-revolution at the points of intersection of their orbits. In the meantime, they recede from one another through a maximal distance $l = r_0\alpha = r_0(\Delta v/v_c)$.

2. The body is ejected from the station in the radial direction, e.g. downward (toward the Earth): $\dot{x}(0) = -\Delta v$, $\dot{y}(0) = 0$ and $\dot{z}(0) = 0$. For these initial conditions, the second equation of (11.43) gives $\dot{y} = -2\,\Omega x$. Substituting $\dot{y}$ into the first equation of (11.43), we obtain $\ddot{x} = -\Omega^2 x$, whence $x(t) = -(\Delta v/\Omega)\sin\Omega t$. Now from the equation $\dot{y} = -2\,\Omega x$ at the initial condition $y(0) = 0$ we find $y(t) = -2(\Delta v/\Omega)(\cos\Omega t - 1)$.

Since $z(t) = 0$, the relative motion of the body described by these equations occurs in the $x$, $y$-plane. To find an explicit expression for the shape of the trajectory, we eliminate the time $t$ from the equations for $x(t)$ and $y(t)$. We obtain

$$\frac{x^2}{l^2} + \frac{(y - 2l)^2}{(2l)^2} = 1. \tag{11.44}$$

Here we again make use of the notation $l = \Delta v/\Omega = r_0(\Delta v/v_c)$ for the characteristic distance. From (11.44) we see that in the reference frame associated with the orbital station, the body moves along an ellipse (figure 11.10($a$)), whose semiminor axis equals $l = r_0(\Delta v/v_c)$ (about 14 km for $\Delta v = 15$ m s$^{-1}$) and whose semimajor axis is twice as long and is oriented along the orbital velocity of the station.

Thus the body ejected downward toward the Earth does in fact at first move vertically down, relative to the station. This is what the astronauts on the station see when they watch the motion of the body with the naked eye. However as the body continues moving, it gradually deviates from its almost vertical initial trajectory. After a half-revolution of the station in its orbit, the body is located in front of the station a distance of $4l$ (about 56 km using the earlier numerical example). After the next half-revolution, the body returns to the station from the opposite side, that is, from above (see figure 11.10($a$)).

The deviation of the body from its initial downward path becomes considerable only after the body has moved several kilometers away from the station. The astronauts will very likely have lost sight of a small body by then. During the period that they can keep the body in sight, it moves toward the ground; when it reappears to the unaided eye, it does so from above!

This almost periodic motion of the body relative to the station can also be easily explained from the point of view of a geocentric observer. Such an explanation is given in section 5.3 together with a description of the corresponding simulation experiment.

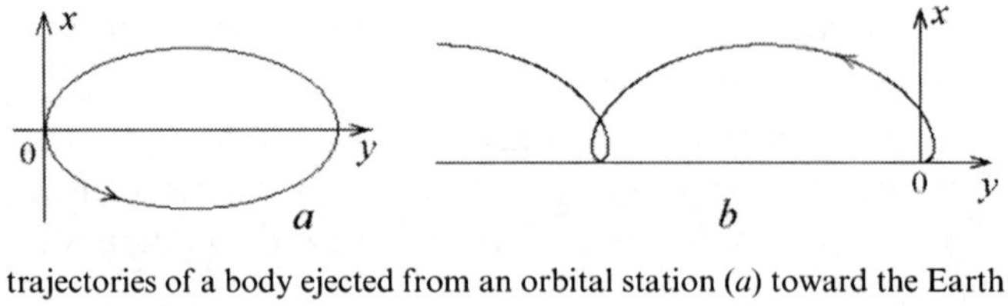


**Figure 11.10.** The trajectories of a body ejected from an orbital station ($a$) toward the Earth and ($b$) forward in the direction of the orbital motion.

3. If the body is ejected from the station parallel or antiparallel to the orbital velocity $\mathbf{v}_c$, its period of revolution along its elliptical orbit is no longer equal to the period of revolution of the station. A *secular term* that steadily increases with time appears in the equations for its relative motion.

For the initial conditions $\dot{x}(0) = 0$, $\dot{y}(0) = \Delta v$ and $\dot{z}(0) = 0$, the particular solution of the system of (11.43) is:

$$x(t) = 2l(1 - \cos \Omega t),$$
$$y(t) = l(-3\,\Omega t + 4 \sin \Omega t), \qquad (11.45)$$
$$z(t) = 0.$$

Here the previous notation $l = r_0(\Delta v/v_c)$ for the characteristic distance is used. In the radial direction (along the local vertical line) the relative motion is again periodic. However, in the direction tangential to the circular orbit (along the $y$-axis), simultaneously with periodic oscillations, a steady (monotone) variation of the coordinate occurs with an average velocity of $-3\Omega l = -3\Delta v$.

The trajectory of this motion is shown in figure 11.10(*b*). (The scale along the ordinate axis is exaggerated.) At first the body, ejected forward from the station, does in fact move forward relative to the station, in the direction of the relative initial velocity imparted to the body. But soon the body turns upward and begins a retrograde motion. After a time $T$ that equals the period of revolution of the station, the body returns to its initial height above the Earth, but behind the station by a distance $6\pi l$ (about 265 km for the quite small initial velocity $\Delta v = 15$ m s$^{-1}$ used in the earlier examples). After the next interval $T$ the distance of the body behind the station doubles, and so on.

A diagram of the relative motion in this case (obtained in a simulation experiment for a rather large value of the initial velocity of the body) is shown in the right panel of figure 5.5.

## 11.8 The gravitational field of a distorted planet

If the distribution of mass in a celestial body is spherically symmetric, the gravitational field outside the body is equivalent to the gravitational field of a point mass, that is, the field is the same as if all the mass were concentrated at the center of the body. The strength of this field decreases with the square of the distance from the center of the body. The distribution of mass in stars and planets is almost spherical and in most problems of celestial mechanics we can consider their gravitational fields as obeying the inverse-square law with great precision.

However, the actual form of a planet can slightly differ from a sphere. For example, the polar radius of the Earth is 21 km smaller than the equatorial radius; that is, the Earth is slightly squeezed along its axis of rotation. One of the programs of the package ('Precession of the Orbit of an Equatorial Satellite') simulates the motion of a satellite in the equatorial plane of such a planet. (For a description of the relevant simulations see chapter 6). Next we show that because of a small axial distortion of a planet, its gravitational field differs from the inverse-square field of a

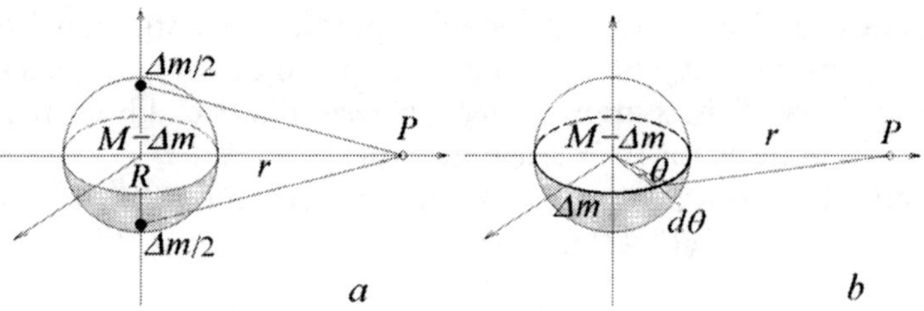


**Figure 11.11.** Models of the mass distribution for a prolate planet: point masses $\Delta m/2$ at the poles (*a*); and for an oblate planet: an equatorial belt of mass $\Delta m$ around the planet (*b*).

point mass by a (small) additional term that falls off with the fourth power of the distance from the planet.

### 11.8.1 A planet with additional masses at the poles

We consider first the case of a prolate planet, whose polar radius is slightly greater than the equatorial one. For large enough distances from the planet, we can assume that its gravitational field is created by an ideal sphere of mass $M - \Delta m$ and two point masses $\Delta m/2$ located at the poles of the planet (figure 11.11(*a*)). In other words, we assume that some small part $\Delta m$ of the whole mass $M$ is transferred from the center through a distance $R$ to the poles of the planet. Because of axial symmetry, the gravitational field in the equatorial plane depends only on the distance $r$ from the center. The potential energy $U(r)$ of a unit mass at some point $P$ in the equatorial plane of this field is the sum of the main term $-G(M - \Delta m)/r$ corresponding to point mass $M - \Delta m$ at the center of the planet, and the term corresponding to two point masses $\Delta m/2$ located at the poles of the planet:

$$U(r) = -\frac{G}{r}(M - \Delta m) - \frac{G\Delta m}{\sqrt{r^2 + R^2}}$$
$$= -\frac{GM}{r}\left(1 - \frac{\Delta m}{M} + \frac{\Delta m}{M}\frac{1}{\sqrt{1 + (R/r)^2}}\right). \tag{11.46}$$

Since this expression for $U(r)$ corresponds to a simplified model of the mass distribution (the point mass $M - \Delta m$ at the center and two point masses $\Delta m/2$ at the poles), it is valid only for distances $r$ from the prolate spheroidal planet that are large compared to the planet's radius $R$: $r \gg R$. But if the radius $R$ of the planet is much smaller than the distance $r$ between the planet and the point $P$ ($R/r \ll 1$), the exact expression for $U(r)$, (11.46), can be simplified and reduced to the following form:

$$U(r) \approx -\frac{GM}{r}\left[1 - \frac{1}{2}\frac{\Delta m}{M}\left(\frac{R}{r}\right)^2\right]. \tag{11.47}$$

The gravitational force $F(r)$ exerted on the unit mass in the equatorial plane can be found as the negative derivative of the potential energy $U(r)$ with respect to $r$:

$$F(r) = -\frac{\mathrm{d}U(r)}{\mathrm{d}r} = -\frac{GM}{r^2}\left[1 - \frac{3}{2}\frac{\Delta m}{M}\left(\frac{R}{r}\right)^2\right]. \tag{11.48}$$

We see that the additional term in the gravitational force $F(r)$ is proportional to the distortion of the planet (characterized by $\Delta m/M$ in our model of the distortion), and inversely proportional to the fourth power of the distance $r$ from the planet. The sign of this term is opposite to that of the main inverse-square term; that is, the axial dilation of the planet reduces the force of gravitation in the equatorial plane. Comparing (11.48) with (6.1) of chapter 6, we obtain the following expression for the dimensionless factor $b$ which is introduced phenomenologically in (6.1) to characterize the additional term of the gravitational force created by an axially dilated planet:

$$b = -\frac{3}{2}\frac{\Delta m}{M}. \tag{11.49}$$

### 11.8.2 A planet with an equatorial bulge

Similarly, we can consider the case of an oblate planet. The gravitational field at large distances from the planet (at $r \gg R$) in this case is the same as that created by an ideally spherical body of mass $M - \Delta m$ with an additional massive belt of mass $\Delta m$ surrounding the planet along the equator (figure 11.11(*b*)). In other words, in our model we assume that some part $\Delta m$ of the total mass $M$ is transferred from the center to the equator of the planet.

To calculate the gravitational field created by this system, we divide the belt into elementary parts subtended by a central angle of $\mathrm{d}\theta$. The mass of each elementary part is $\mathrm{d}m = (\Delta m/2\pi)\mathrm{d}\theta$. If the angular position of an elementary part of the belt is characterized by an angle $\theta$ (see figure 11.11(*b*)), its distance from point $P$ equals $\sqrt{r^2 + R^2 - 2rR\cos\theta}$ and its contribution $\mathrm{d}U(r)$ in the potential energy of a unit mass at point $P$ is

$$\mathrm{d}U(r) = -\frac{G\mathrm{d}m}{\sqrt{r^2 + R^2 - 2rR\cos\theta}} = -\frac{G\Delta m}{2\pi}\frac{\mathrm{d}\theta}{\sqrt{r^2 + R^2 - 2rR\cos\theta}}. \tag{11.50}$$

To find the potential energy $\Delta U(r)$ of the unit mass in the field created by the entire belt, we use the principle of superposition and integrate $\mathrm{d}U(r)$, (11.50), over $\theta$ along the belt from 0 to $2\pi$:

$$\begin{aligned}\Delta U(r) &= -\frac{G\Delta m}{2\pi}\int_0^{2\pi}\frac{\mathrm{d}\theta}{\sqrt{r^2 + R^2 - 2rR\cos\theta}}\\ &= \frac{G\Delta m}{r}\frac{1}{2\pi}\int_0^{2\pi}\frac{\mathrm{d}\theta}{\sqrt{1 + (R/r)^2 - 2(R/r)\cos\theta}}.\end{aligned} \tag{11.51}$$

Our model of the mass distribution for an oblate planet is appropriate only for large distances $r \gg R$, so we can simplify the integrand for the case $R/r \ll 1$:

$$\Delta U(r) \approx -\frac{G\Delta m}{r}\frac{1}{2\pi}\int_0^{2\pi}\left[1 - \frac{1}{2}\left(\frac{R}{r}\right)^2 + \frac{R}{r}\cos\theta + \frac{3}{2}\left(\frac{R}{r}\right)^2\cos^2\theta\right]d\theta$$

$$= -\frac{G\Delta m}{r}\left[1 + \frac{1}{2}\left(\frac{R}{r}\right)^2\frac{1}{2\pi}\int_0^{2\pi}(3\cos^2\theta - 1)d\theta\right]$$

$$= -\frac{G\Delta m}{r}\left[1 + \frac{1}{4}\left(\frac{R}{r}\right)^2\right]. \qquad (11.52)$$

Adding the potential energy $\Delta U(r)$ created by the belt to the potential energy $-G(M - \Delta m)/r$ of the unit mass in the gravitational field created by the point mass $M - \Delta m$ located at the center of the planet, we obtain the following expression for the potential energy at a distance $r$ in the gravitational field of the oblate planet:

$$U(r) = -\frac{GM}{r}\left[1 + \frac{1}{4}\frac{\Delta m}{M}\left(\frac{R}{r}\right)^2\right] = -\frac{gR^2}{r}\left[1 + \frac{1}{4}\frac{\Delta m}{M}\left(\frac{R}{r}\right)^2\right]. \qquad (11.53)$$

We replaced here the product $GM$ with $gR^2$, where $g$ is the acceleration of free-fall at the surface of the planet. Differentiating the potential energy $U(r)$ with respect to $r$, we find the gravitational force exerted on the unit mass in the equatorial plane of the oblate planet:

$$F(r) = -\frac{dU(r)}{dr} = -\frac{gR^2}{r^2}\left[1 + \frac{3}{4}\frac{\Delta m}{M}\left(\frac{R}{r}\right)^2\right] = -\frac{gR^2}{r^2}\left[1 + b\left(\frac{R}{r}\right)^2\right]. \qquad (11.54)$$

Here we have defined a dimensionless parameter $b = (3/4)(\Delta m/M)$ to characterize the axial distortion of the planet. This parameter was earlier introduced phenomenologically (see (6.1) in chapter 6).

Thus our model of an oblate planet as a sphere with an equatorial belt produces in $F(r)$ an additional term that is proportional to the distortion of the planet. This term decreases as the fourth power of the distance $r$ from the planet and its sign is the same as that of the main (inverse-square) term; that is, the axial contraction of the planet increases the gravitational force in the equatorial plane.

## 11.9 The two-body problem

The two-body problem is concerned with the motion of two interacting bodies (point masses) whose masses are $m_1$ and $m_2$. If the mass of one of the bodies is much greater than that of the other, the heavier body can be treated as stationary. In other words, we can fix a reference frame to the heavier body and this reference frame can be considered with great accuracy to be an inertial one. Thus the problem is reduced to the study of the motion of the lighter body in this inertial reference frame. This body moves in a given stationary force field, which is created by the heavier body. Such a situation can be called a *one-body problem*.

However, this approximation is not applicable if the masses of the two interacting bodies are comparable. For instance, if the components of a double star have almost equal masses, neither of the components may be treated as stationary. Since it is necessary to take into account the motion of both interacting bodies, we are obliged to deal with a *two-body problem*.

### 11.9.1 Reduced mass and relative motion

In most textbooks on mechanics, the two-body system is generally not referred to as a many-body system because the problem of the relative motion of two interacting bodies (irrespective of the physical nature of the interaction) is mathematically equivalent to the problem of motion of a virtual single body with the reduced mass $\mu = m_1 m_2/(m_1 + m_2)$ under a stationary central force equal to the force of interaction between the actual bodies. The proof is given below. The solution to this problem actually describes the motion of one body relative to the other. Under the inverse-square gravitational force of interaction, this relative motion of the bodies occurs along a conic section and obeys Kepler's laws. Knowing that if one body moves, say, in an ellipse about the other (a binary star), we can show that they both move synchronously in homothetic ellipses about the center of mass of the whole system. As the bodies move, they are always at the ends of a rotating straight line that passes through the common focus of their orbits located at the stationary center of mass (see figure 7.1). The linear dimensions of these similar elliptical orbits are inversely proportional to the masses of the bodies.

We let the positions of the two bodies in a given inertial frame be determined by radius vectors $\mathbf{r}_1$ and $\mathbf{r}_2$. We denote by $\mathbf{F}_{12}$ the central force with which the second body of mass $m_2$ acts upon the first body of mass $m_1$. Then, in accordance with Newton's third law, the first body acts on the second with a force $\mathbf{F}_{21}$ of equal magnitude and opposite direction:

$$\mathbf{F}_{21} = -\mathbf{F}_{12}. \qquad (11.55)$$

It is not necessary to specify here the physical nature of the interaction between the two bodies, but in particular, it can be the gravitational force. Next we write the differential equations of motion, Newton's second law, for both bodies:

$$m_1\ddot{\mathbf{r}}_1 = \mathbf{F}_{12} \qquad m_2\ddot{\mathbf{r}}_2 = -\mathbf{F}_{12}. \qquad (11.56)$$

Next we transform these equations from the variables $\mathbf{r}_1$ and $\mathbf{r}_2$ to new variables $\mathbf{r}_c$ and $\mathbf{r}$, which are related to $\mathbf{r}_1$ and $\mathbf{r}_2$ by:

$$\mathbf{r}_c = \frac{m_1\mathbf{r}_1 + m_2\mathbf{r}_2}{m_1 + m_2} \qquad \mathbf{r} = \mathbf{r}_1 - \mathbf{r}_2. \qquad (11.57)$$

The new variables $\mathbf{r}_c$ and $\mathbf{r}$ have a definite physical meaning: the radius vector $\mathbf{r}_c$ is the displacement of the center of mass of the two-body system relative to the origin of the inertial frame and the vector $\mathbf{r}$ is the displacement of the first body relative to the second. Adding the equations of motion, we obtain $\ddot{\mathbf{r}}_c = 0$: the center of mass of the system of two interacting bodies moves in a straight line at constant speed. We emphasize that the acceleration of the center of mass is zero because the forces of interaction satisfy Newton's third law, expressed by (11.55).[1]

To obtain the differential equation for the variable $\mathbf{r}$ that describes the relative motion of the bodies, we divide the first of these equations, (11.56), by $m_1$, the second equation by $m_2$, and subtract the second equation from the first. We thus obtain:

$$\ddot{\mathbf{r}} = \left(\frac{1}{m_1} + \frac{1}{m_2}\right)\mathbf{F}_{12}. \tag{11.58}$$

We next introduce the *reduced mass* $\mu$ of the system by the definition:

$$\frac{1}{\mu} = \frac{1}{m_1} + \frac{1}{m_2} \qquad \text{or} \qquad \mu = \frac{m_1 m_2}{m_1 + m_2}. \tag{11.59}$$

Then (11.58) acquires the sense of an equation of motion of a *single body* of mass $\mu$ under the action of the force $\mathbf{F}_{12}$, which remains the original force of interaction between the bodies. In particular, when the interaction of the bodies is caused by universal gravitation, the problem of determining their relative motion (of finding the function $\mathbf{r}(t)$ that determines the position of the first body relative to the second) is reduced to Kepler's problem for a single body of mass $\mu$ in a central field of gravitation under the action of a force that is inversely proportional to the square of distance from the center of force. Solving this problem, we find the motion of one of the interacting bodies relative to the other. Having done so, we can easily find the motion of both bodies in the original inertial frame.

It is convenient to consider the motion of each of the bodies in an inertial frame fixed to the center of mass of the system of bodies. We choose the origin at the center of mass, assuming $\mathbf{r}_c = 0$, that is, $m_2\mathbf{r}_2 = -m_1\mathbf{r}_1$. Then the radius vector of each of the bodies is related to the radius vector $\mathbf{r}(t)$ of the relative position by:

$$\mathbf{r}_1 = \frac{m_1}{m_1 + m_2}\mathbf{r} \qquad \mathbf{r}_2 = -\frac{m_2}{m_1 + m_2}\mathbf{r}. \tag{11.60}$$

This means that in the case in which the relative motion described by (11.58) occurs along some Keplerian ellipse under the action of gravitational forces, both bodies move along geometrically similar (homothetic) Keplerian orbits whose common focus (homothetic center) is located at the center of mass of the system (see chapter 7, figure 7.1). The bodies in their motion remain at the opposite sides of

[1] However, one can consider the conservation of momentum of an isolated system, as well as the uniform motion of its center of mass (which is related with this conservation), to be a more fundamental property which reflects the homogeneity of physical space, and regard Newton's third law (and the requirement of its fulfillment for all fundamental interactions, including the law of universal gravitation) as a consequence of the conservation of momentum.

a rectilinear segment that passes through the center of mass of the system. The linear dimensions of these homothetic orbits are inversely proportional to the masses of the bodies.

If mass $m$ of one of the bodies is much less than mass $M$ of the other (which occurs, for example, in the case of a planet orbiting the Sun, or in the case of a satellite orbiting a planet), the center of mass very nearly coincides with the position of the massive body and the reduced mass almost equals the mass of the lighter body. Thus, in this case we return to the problem of motion of a body under the force of attraction toward a stationary center of force. In the case in which the masses of the interacting bodies are equal, the reduced mass is one half the mass of one of the bodies. Here the bodies move synchronously in equal (congruent) orbits and the relative motion of the bodies occurs along a geometrically similar (homothetic) orbit whose size is twice the size of the orbits traced by the bodies in the center-of-mass frame.

### 11.9.2 An alternative approach to the two-body problem

As shown above, the concept of the reduced mass enables us to regard the two-body problem as a one-body problem. This traditional approach, being quite correct and mathematically simple, may appear especially amazing and cause confusion since it allows us to treat the non-inertial frame of either of the bodies as inertial. The explanation of this apparent inconsistency is all too subtle for most students who study physics at an introductory level. Moreover, the business of transforming from one frame to another in this case can also be rather confusing. (After all, both Copernicus and Galileo had difficulty getting the world to accept such ideas.)

However, dealing with the two-body problem, it is possible to use a somewhat different approach which is free of the difficulties mentioned. We consider the motion of each body in the inertial center-of-mass frame of reference. Since the force of gravity between the bodies lies at each instant along the line joining the bodies, the force vectors are always directed through the center of mass. In order to explain why the motions of the bodies relative to the center of mass obey Kepler's laws and occur along conic sections, it is sufficient to show that each of the bodies coupled by mutual gravitation can be treated as moving not under the pull of the other moving body, but rather in a stationary central gravitational field whose strength diminishes with the square of the distance from the center of mass. The source of the field (located at the stationary center of mass) is characterized by some effective mass $M_{\text{eff}}$.

Indeed, let $\mathbf{r}_1$ and $\mathbf{r}_2$ be the radius vectors denoting momentary positions relative to the center of mass of the bodies with masses $m_1$ and $m_2$, respectively. Then $m_1\mathbf{r}_1 + m_2\mathbf{r}_2 = 0$ and $r_1 + r_2 = [1 + (m_1/m_2)]r_1$. Therefore, in the formula for the gravitational force $\mathbf{F}_{12}$ exerted on the first body, we can express the distance $(r_1 + r_2)$ between the bodies in terms of the sole distance $r_1$ of the first body from the center of mass:

$$\mathbf{F}_{12} = -G\frac{m_1 m_2}{(r_1 + r_2)^2}\frac{\mathbf{r}_1}{r_1} = -G\frac{m_1 (M_1)_{\text{eff}}}{r_1^2}\frac{\mathbf{r}_1}{r_1}, \qquad (M_1)_{\text{eff}} = \frac{m_2^3}{(m_1 + m_2)^2}. \tag{11.61}$$

Thus, in the inertial center-of-mass frame, the motion of the first body in the two-body system should be exactly the same as in a central inverse-square gravitational field created by a stationary source of mass $(M_1)_{\text{eff}}$ (in the absence of the second body). As we know, such a motion obeys Kepler's laws.

Similar considerations can be applied to the other body of the two-body system: the gravitational influence of the first body can be replaced by the stationary source of an effective mass $(M_2)_{\text{eff}} = m_1^3/(m_1 + m_2)^2$. It remains only to prove that these Keplerian motions of both bodies occur synchronously along homothetic (closed or open) coplanar orbits whose linear dimensions are inversely proportional to the masses of the bodies. This follows immediately from the equation $m_1\mathbf{r}_1 + m_2\mathbf{r}_2 = 0$, which holds for an arbitrary moment during the motion.

In another inertial reference frame (that is, as seen from outside) the bodies of a two-body system move non-uniformly along complicated wavy or looped trajectories. This apparent complexity is generated by the superposition of their rather simple periodic Keplerian elliptical motions around the center of mass and quite simple uniform rectilinear motion along the center of mass. Figure 7.2 shows an example of such trajectories.

## 11.10 Exact particular solutions to the three-body problem

The three-body problem is frequently mentioned in various intermediate mechanics texts as an example of the extreme complexity of possible motions generated by simple and precise physical laws. However, since Lagrange it has been well known that in generally unsolvable many-body problems there exist several particular solutions describing simple Keplerian motions of the bodies. It may seem a real wonder that such an unexpectedly simple finite subset of motions falls out of the continuous set of tremendously complex general three-body motions. These simple solutions should certainly allow an equally simple physical explanation.

In all cases in which the motion of individual bodies (coupled by mutual gravitational forces in a many-body system) occurs along conic sections, each body can be treated as moving not under the pull of the other moving bodies, but rather in a stationary central gravitational field whose strength is inversely proportional to the square of the distance of the body from the center of mass of the system. We note that under certain conditions this effective gravitational field can be stationary in spite of the fact that it is created by moving bodies.

For the *circular* restricted three-body problem, there exists a class of exact solutions corresponding to the equilibrium of the third body (whose mass is negligible compared to masses of the other two bodies) at one of the five *libration points* (or Lagrange points) in the reference frame rotating with the line joining the massive bodies. Simulations of the corresponding motions are described in section 8.3 (see also the simulation program 'Planet with a Satellite'). Below we consider a mathematical proof of the existence of the triangular libration points and describe a method to calculate the positions of the collinear libration points.

The existence of the triangular libration points is nearly obvious in the special case in which the masses of the bodies are equal and in which (in the center-of-mass

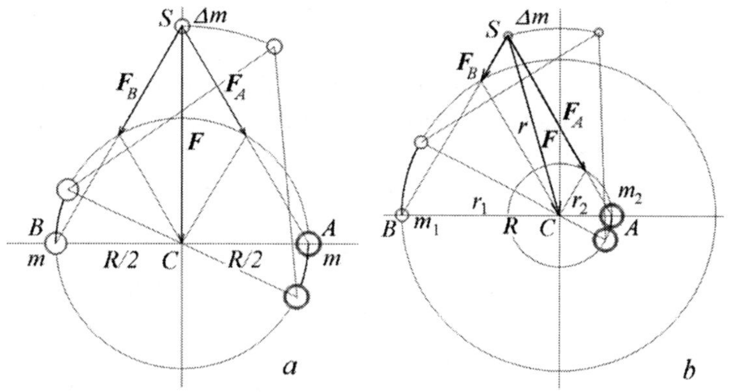


**Figure 11.12.** The triangular libration points in a system with equal masses of the bodies *A* and *B* (*a*) and with different masses (*b*).

frame) the bodies move opposite one another around a common circle whose diameter is *R*. (see figure 11.12(*a*).) The angular velocity $\omega$ of their rotation can be calculated using Newton's second law. Applying the law to one of the massive bodies (of mass *m*) that moves along a circle of radius $R/2$ under the force of gravitational attraction by the other body, we write:

$$m\omega^2 R/2 = Gm^2/R^2,$$

whence $\omega^2 = 2Gm/R^3$. If we place the third body (of a negligible mass $\Delta m$) at the vertex *S* of the equilateral triangle *BAS* whose base *BA* is the segment joining the massive bodies, the resulting force of gravitational attraction on the third body by the two massive bodies is directed toward the center of mass *C* and its magnitude has the value needed to provide the third body with the centripetal acceleration necessary for a circular motion synchronous with that of the massive bodies. Indeed, the net gravitational force equals $\sqrt{3}Gm\Delta m/R^2$, while the centripetal acceleration of rotation (about the center of mass *C* along the circle of radius $R\sqrt{3}/2$ with the angular velocity $\omega$) equals $\omega^2 R\sqrt{3}/2 = \sqrt{3}Gm/R^2$.

Thus, if the third body at the libration point *S* is initially at rest in the rotating reference frame, it remains there indefinitely. In other words, the whole system of three bodies rotates rigidly about the center of mass *C* with the angular velocity $\omega = \sqrt{2mG/R^3}$. The triangular equilateral configuration of the system is preserved during the motion. In the reference frame associated with one of the massive bodies, say *A*, the third body moves around *A* along the same circular orbit as does the other massive body *B*, either behind or in front of *B* by an angle of 60°.

For a system in which the masses of the two massive bodies are different, the argument is a bit more complicated. The circular motion of the system about the center of mass *C* (figure 11.12(*b*)) occurs at an angular velocity $\omega = \sqrt{(m_1 + m_2)G/R^3}$. This expression is found in the same way as in the preceding case of equal masses, by applying Newton's second law to the circular motion of one of the bodies. To prove that in this case the system of the three bodies can also rotate rigidly, we consider the gravitational forces exerted on the third body by the massive bodies. We must show that the resulting force is directed toward the center of mass and its magnitude has exactly the value needed to provide the third body with the necessary centripetal acceleration.

We draw a vector from the triangular libration point $S$ (figure 11.12($b$)) to the center of mass $C$ and consider its components along the lateral sides of the triangle $BAS$ (along the directions from $S$ to $A$ and $B$). It is clear from the figure that these components are equal to the distances $r_1$ and $r_2$ between the center of mass and masses $B$ and $A$, respectively. Next we show that this vector can be treated as the resulting gravitational force $\mathbf{F}$ (in some scale) exerted on the third body $S$ of a negligible mass $\Delta m$, and its components along the sides of the triangle as the individual gravitational forces $\mathbf{F}_B$ and $\mathbf{F}_A$ exerted on $S$ by the body $B$ of mass $m_1$ and the body $A$ of mass $m_2$, respectively.

We let $R$ be the distance between the massive bodies $A$ and $B$. Then $r_1 = Rm_2/(m_1 + m_2)$ and $r_2 = Rm_1/(m_1 + m_2)$. According to the law of gravitation, $F_B = Gm_1\Delta m/R^2$ and $F_A = Gm_2\Delta m/R^2$, because the distances between the libration point $S$ and each of the bodies $A$ and $B$ also equal $R$. We see that the ratio $F_A/F_B$ equals $m_2/m_1 = r_1/r_2$, and hence the components of the vector $SC$ can indeed be treated as gravitational forces $\mathbf{F}_A$ and $\mathbf{F}_B$ exerted on $S$ by the bodies $A$ and $B$. Therefore the sum of $\mathbf{F}_A$ and $\mathbf{F}_B$ is actually directed toward the center of mass $C$.

It remains only to show that the magnitude of the net force $\mathbf{F} = \mathbf{F}_A + \mathbf{F}_B$ has the value that is required in order to provide the third body of mass $\Delta m$ with the necessary centripetal acceleration $\omega^2 r = (m_1 + m_2)Gr/R^3$ (in the inertial, center-of-mass frame of reference), where $r$ is the distance between the center of mass $C$ and the triangular libration point $S$. From figure 11.12($b$) we see that $F/F_B = r/r_2$, and, consequently, $F/F_B = (r/R)(m_1 + m_2)/m_1$. Since $F_B = Gm_1\Delta m/R^2$, we find that $F = Gr/R^3(m_1 + m_2)\Delta m$; that is, thc force has exactly the required value.

Thus, we have proved that the system can rotate rigidly about the center of mass. The bodies move synchronously along concentric circles of different radii. In the reference frame associated with one of the massive bodies, the other two bodies move around it along the same circular orbit at an angular distance of 60°. At one of the triangular libration points the light body lags behind the massive body by 60°, at the other libration point the light body leads the massive body by the same angle.

The stability in the motion in the vicinity of the Lagrange triangular libration points in the restricted planar circular three-body problem remained a subject of intense investigation in celestial mechanics for more than two centuries. It was found that the triangular libration points are stable for the mass ratio $\kappa = m_1/(m_1 + m_2)$ (where $m_1 < m_2$), satisfying the following condition:

$$\kappa(1 - \kappa) < 1/27.$$

That is, the triangular libration points are stable if the mass of one of the massive bodies is much smaller than that of the other (if the ratio $m_1/m_2$ does not exceed

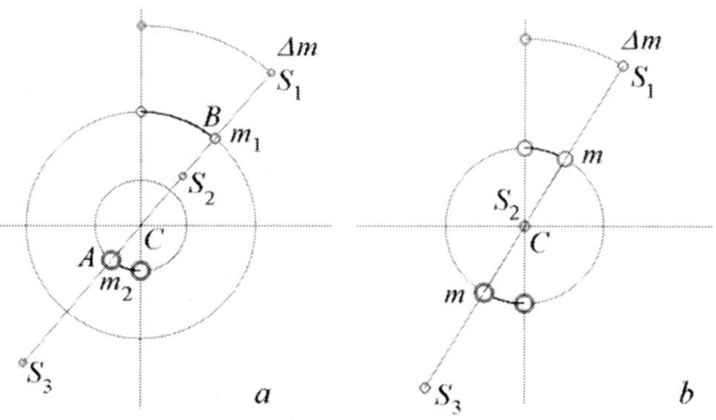


**Figure 11.13.** The collinear libration points $S_1$, $S_2$ and $S_3$ in a system with different masses of the bodies $A$ and $B$ ($a$) and with equal masses ($b$).

approximately 0.04). In the Earth–Moon system $m_1/m_2 = 0.0123$ and so its triangular libration points are stable.

Next we determine the position of the three collinear libration points $S_1$, $S_2$ and $S_3$ (figure 11.13($a$)). If the mass of one of the massive bodies is considerably smaller than the mass of the other, two of the collinear libration points occur near the smaller body ($B$ in figure 11.13($a$)). It is convenient then to regard the third body (of a negligible mass) at one of these points as a satellite of the smaller body, in spite of the fact that the third body at a libration point is actually a satellite of both massive bodies $A$ and $B$ simultaneously. Such a satellite moves synchronously with $A$ and $B$ in their relative revolution. This revolution occurs with an angular velocity $\omega$ that depends on the total mass $m_1 + m_2$ of the system and on the distance $R$ between the massive bodies: $\omega = \sqrt{(m_1 + m_2)G/R^3}$.

To determine the position of any of these points, say, the distance $l_1$ of the outer libration point $S_1$ from the center of mass $C$, we can use the law of gravitation and Newton's second law. The resulting force exerted on the satellite at point $S_1$ is the vector sum of the gravitational forces created by the massive bodies $B$ and $A$. Both these forces are directed toward the center of mass $C$, and their sum must provide the satellite with a centripetal acceleration $\omega^2 l_1$ that is required for the circular motion about the center of mass with the given angular velocity $\omega$. Therefore,

$$\omega^2 l_1 = G\left[\frac{m_1}{(l_1 - r_1)^2} + \frac{m_2}{(l_1 + r_2)^2}\right]. \tag{11.62}$$

Here $r_1$ and $r_2$ are the distances between the center of mass $C$ of the system and the massive bodies $B$ (of mass $m_1$) and $A$ (of mass $m_2$) respectively: $r_1 = Rm_2/(m_1 + m_2)$, $r_2 = Rm_1/(m_1 + m_2)$. The first term in the right side of (11.62) is due to the gravitational attraction to body $B$ whose distance from $S_1$ equals $l_1 - r_1$, and the second term is due to the gravitational attraction to body $A$ whose distance is $l_1 + r_2$.

Substituting into (11.62) the given value of the angular velocity squared, $\omega^2 = (m_1 + m_2)G/R^3$, we obtain an equation for determining the distance $l_1$:

$$(m_1 + m_2)\frac{l_1}{R^3} = \frac{m_1}{(l_1 - r_1)^2} + \frac{m_2}{(l_1 + r_2)^2}. \tag{11.63}$$

This equation can be solved numerically by iterations. First we rewrite (11.63) to give it the form $x = f(x)$ with $f'(x) < 1$:

$$l_1 = r_1 + \sqrt{\frac{m_1}{(m_1 + m_2)l_1/R^3 - m_2/(l_1 + r_2)^2}}. \tag{11.64}$$

Taking $l_1 = R$ as the first approximation, we substitute it into the right side of (11.64). The left side is then the second approximation for $l_1$. We substitute this approximate value of $l_1$ into the right side, and obtain the next approximation for $l_1$; and so on, until an approximate value agrees with the preceding one to within a satisfactory level of accuracy.

In this way we find that the distance $l_1$ of the outer libration point from the center of mass for the system with $m_1 = m_2$ equals 1.19841 $R$; for $m_1 = 0.5\, m_2$ it equals 1.24905 $R$; for $m_1 = 0.4\, m_2$ $l_1 = 1.25967\, R$. For the Earth–Moon system $m_1 = 0.0123\, m_2$ and so the distance of the outer libration point from the center of mass is 1.155 68 times the mean distance $R$ between the Earth and the Moon, or approximately 0.17 $R$ from the Moon.

Similarly we can find the position of the inner collinear libration point $S_2$. Applying Newton's second law to the circular motion of a satellite at this point, we take into account that the gravitational pull from body $B$ is directed away from the center of revolution. For the distance $l_2$ of this libration point from the center of mass $C$, we obtain the following values: in the system with equal masses of the two bodies ($m_1 = m_2$), this point occurs at the center of mass, that is, $l_2 = 0$ (evident from symmetry considerations); for the system with $m_1 = 0.5\, m_2$ $l_2 = 0.237\,418\, R$; for $m_1 = 0.4\, m_2$ $l_2 = 0.307\,233\, R$. For the Earth–Moon system $l_2 = 0.836\,915\, R$, or approximately at a distance 0.15 $R$ from the Moon.

The third collinear libration point $S_3$ lies on the opposite side (with respect to $B$) of the larger body $A$. In the system in which the masses of $A$ and $B$ are equal (figure 11.13(*b*)) its distance from the center of mass is the same as for the libration point $S_1$: $l_3 = l_1 = 1.19841\, R$. This equality is evident from the symmetry of the system. For the system with $m_1 = 0.5\, m_2$ $l_3 = 1.136\,36\, R$; for $m_1 = 0.4\, m_2$ $l_3 = 1.11\,751\, R$. For the Earth–Moon system $l_3 = 1.00506\, R$, or approximately 0.993 $R$ on the side of the Earth opposite the Moon. A satellite at this libration point moves around the Earth along almost the same circular orbit as does the Moon, provided the initial velocity of the satellite has a proper value. During the motion, the Moon and the satellite are located at the diametrically opposite points of the orbit.

We discussed above exact particular solutions to the restricted three-body problem that describe circular motions. However, similar exact solutions associated with the libration points also exist for elliptic (and hyperbolic) motions of the three bodies. If in the system of two massive bodies that revolve about the center of mass in homothetic elliptical orbits, a body of a negligible mass (a satellite) is placed at one of the three collinear libration points, this body can also synchronously trace a homothetic elliptical orbit provided its initial velocity has a proper value. During the motion all three bodies remain on the same straight line. Although in this motion the distances between the bodies vary, the ratio of these distances remains the same. An example of such motion at an inner libration point is shown in figure 8.13.

To explain the observed motions, we should first take into account that the sum of the forces exerted on the satellite at a collinear libration point by both massive bodies is inversely proportional to the square of the distance between the center of mass and the satellite. If the collinear configuration with the constant ratio of the distances between the bodies is preserved during the motion, we can consider the motion of the satellite to occur in a *stationary* effective central gravitational field whose source is located at the center of mass. Next we should show that this effective field at any of the collinear libration points gives the satellite an acceleration (toward the center of mass) that relates to the acceleration of a massive body (under the gravitational pull of the other body) just as the distance of the satellite to the center of mass relates to the distance of the corresponding massive body to the center of mass. Then, if the initial velocity of the satellite is related in the same way to the velocity of the massive body, the satellite and the body trace synchronously homothetic Keplerian orbits. The collinear configuration is preserved during the motion.

Elliptic (and hyperbolic) motions are also possible if a satellite is located at a triangular libration point (at the third vertex of the equilateral triangle whose base is the line joining the massive bodies). The equilateral triangular configuration of the system is preserved during the motion in which the three bodies move synchronously in ellipses with a common focus at the center of mass of the system. An example of this motion is shown in figure 8.12. The triangle $ABS$ formed by the bodies rotates non-uniformly and the lengths of its sides vary periodically during the rotation. That this periodic Keplerian motion of the satellite is possible can be similarly shown using the concept of a stationary effective central gravitational field with a source at the center of mass. The resulting force of attraction by the two massive bodies, as we have shown above for the case of circular motions, is again directed to the center of mass and its magnitude is inversely proportional to the square of the distance between the satellite and the center of mass.

## 11.11 The non-restricted three-body problem

The exact particular solutions in which the bodies synchronously trace congruent elliptical (or hyperbolic) orbits lying in the same plane exist even for the *unrestricted* three-body problem in which the mass of one of the bodies cannot be considered as negligible. The existence of such simple particular solutions was first shown by Euler and Lagrange. Two types of such solutions correspond to equilateral triangular configurations of the bodies and three more types to collinear configurations. For each of the three possible collinear configurations, the distances between the bodies should be in a certain ratio that is preserved during the motion. This ratio depends on the masses $m_1$, $m_2$ and $m_3$ of the bodies and is determined by an equation of the 5th power. Even though the practical importance of these particular solutions may seem doubtful, their existence is curious in principle.

### 11.11.1 A star with two identical planets

The possibility of simple Keplerian motions in the collinear configuration is almost evident for the special case of a system of two planets of equal mass that orbit a single star. Let the planets initially lie on the same straight line with the star at the mid point (figure 11.14). If the planets have equal and opposite initial velocities (in the frame of reference of the star), this configuration of the three bodies is preserved during their further motion.

Next we show that the motion of the planets in this system is Keplerian. The center of mass of the system is initially located at the center of the star and it remains there during the motion since the planets have equal and opposite velocities. The net force exerted on each of the planets is formed by the addition of the forces of gravitational attraction to the star and to the other planet. This resulting force is always directed toward the center of the star and its magnitude is inversely proportional to the square of the distance from the star. Thus

$$F = G\frac{mM}{r^2} + G\frac{mm}{(2r)^2} = G\frac{m(M + m/4)}{r^2}. \tag{11.65}$$

Here $G$ is the gravitational constant, $M$ is the mass of the star, $m$ is the mass of either of the planets and $r$ is the distance from the star to either of the planets.

It follows from (11.65) that in the symmetrical configuration, the motion of each of the planets occurs along a Keplerian ellipse, as if the motion were governed solely by an effective stationary central Newtonian gravitational field whose stationary source is characterized by a mass of $M + m/4$. The planets in such a system move synchronously along identical ellipses with a common focus at the center of mass of the system. At any moment the planets are located at opposite ends of the straight line that passes through the center of the star (see figure 11.14).

With the help of Newton's second law, we can calculate the velocity of the planets for the special case of circular orbits. Equating the force given by (11.65) to the

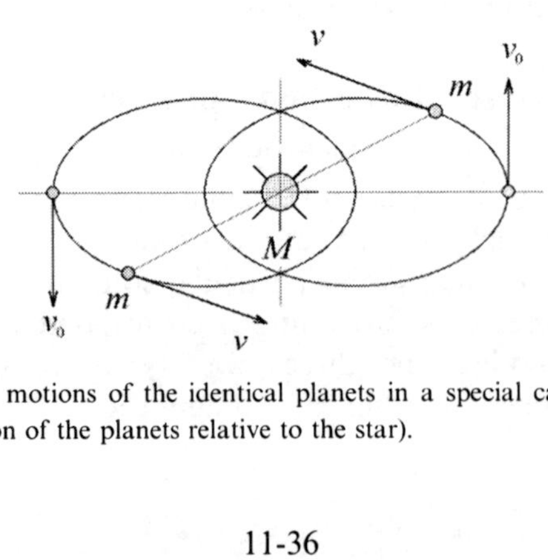


**Figure 11.14.** The Keplerian motions of the identical planets in a special case of the three-body problem (a symmetric collinear position of the planets relative to the star).

product of the planet's mass $m$ and the centripetal acceleration $v_c^2/r$, we obtain for the velocity $v_c$ of the planet in the circular orbit of a radius $r$:

$$v_c = \sqrt{\frac{G}{r}\left(M + \frac{m}{4}\right)}. \tag{11.66}$$

The period of revolution of the planets along such circular orbits is found by dividing the length of the orbit $2\pi r$ by the circular velocity $v_c$:

$$T = \frac{2\pi r}{v_c} = 2\pi\sqrt{\frac{r^3}{G(M + m/4)}}. \tag{11.67}$$

This expression is a generalization of Kepler's third law for the special case of planetary motion under consideration. Equation (11.67) is equally valid for elliptic motions of the planets provided we replace the radius $r$ with the semimajor axis $a$ of the elliptical orbit.

The collinear particular solution of the three-body problem discussed above is also valid for a planet in a double-star system whose components have equal masses. The identical components of the double star move along equal elliptical (or circular) orbits, and the planet that is placed at rest at the center of mass of the system remains there in equilibrium during the motion of the components. In this case in (11.65)–(11.67) we let the mass $m$ be the mass of each of the double-star components and the mass $M$ be the mass of the planet.

The equilibrium of the planet at the center of the segment joining the orbiting components of the double star is unstable. Hence the particular solution of the three-body problem under discussion describes a motion that is unstable with respect to any (arbitrarily small) variations of the initial conditions. We can easily observe this instability in the simulation experiment.

### 11.11.2 The regular Keplerian motions of three different bodies in the equilateral configuration

Next we give a detailed explanation of the exact particular solutions to the unrestricted three-body problem for the triangular equilateral configuration of the bodies. The case in which the masses of the bodies are equal is discussed in section 9.5.3 (see figure 9.14). Below we show analytically that simple Keplerian motions are possible even when the masses of the bodies are different.

Let the bodies *1*, *2* and *3* (of masses $m_1$, $m_2$ and $m_3$, respectively) be located at the vertices of the equilateral triangle whose sides equal $R$ (figure 11.15 shows the system with $m_1 = 0.3m_3$ and $m_2 = 0.6m_3$). We denote by $\mathbf{r}_{12}$ and $\mathbf{r}_{13}$ the radius vectors of bodies *2* and *3* relative to *1* (that is, the vectors joining *1* with *2* and *3*, respectively) and by $\mathbf{F}_{12}$ and $\mathbf{F}_{13}$ the gravitational forces exerted on *1* by *2* and *3*, respectively. According to the law of gravitation,

$$\mathbf{F}_{12} = Gm_1m_2\frac{\mathbf{r}_{12}}{R^3}, \qquad \mathbf{F}_{13} = Gm_1m_3\frac{\mathbf{r}_{13}}{R^3}.$$

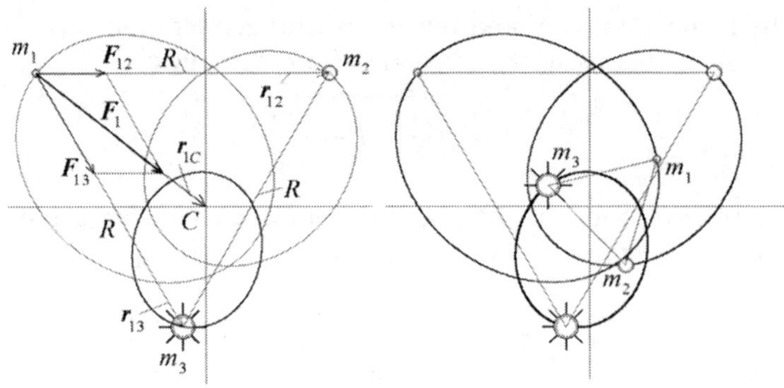


**Figure 11.15.** The regular Keplerian motions of three bodies of different masses in the equilateral configuration.

We add $\mathbf{F}_{12}$ and $\mathbf{F}_{13}$ vectorially to find the total gravitational force $\mathbf{F}_1$ exerted on *1*:

$$\mathbf{F}_1 = \mathbf{F}_{12} + \mathbf{F}_{13} = \frac{Gm_1}{R^3}(m_2\mathbf{r}_{12} + m_3\mathbf{r}_{13}). \tag{11.68}$$

This force $\mathbf{F}_1$ is directed to the center of mass $C$ of the system. Indeed, the radius vector $\mathbf{r}_{1C}$ of the center of mass relative to *1* (the vector joining *1* with $C$) is given by the following expression:

$$\mathbf{r}_{1C} = \frac{(m_2\mathbf{r}_{12} + m_3\mathbf{r}_{13})}{M}, \tag{11.69}$$

where $M = m_1 + m_2 + m_3$ is the total mass of the system. With the help of (11.69), we can express the total force $\mathbf{F}_1$ exerted on the body *1* by the other two bodies *2* and *3* in terms of $M$ and $\mathbf{r}_{1C}$:

$$\mathbf{F}_1 = \mathbf{F}_{12} + \mathbf{F}_{13} = \frac{GMm_1}{R^3}\mathbf{r}_{1C}. \tag{11.70}$$

We conclude from (11.70) that the acceleration $\mathbf{a}_1$ of the body *1* produced by the combined gravitation of bodies *2* and *3* is proportional to $\mathbf{r}_{1C}$. It is clear from symmetry that similar expressions are valid for the accelerations of the other two bodies *2* and *3* of the system:

$$\mathbf{a}_1 = \frac{GM}{R^3}\mathbf{r}_{1C}, \qquad \mathbf{a}_2 = \frac{GM}{R^3}\mathbf{r}_{2C}, \qquad \mathbf{a}_3 = \frac{GM}{R^3}\mathbf{r}_{3C}. \tag{11.71}$$

Here $\mathbf{r}_{2C}$ and $\mathbf{r}_{3C}$ are the vectors joining the bodies *1* and *2* with the center of mass $C$. Therefore the accelerations of all three bodies are directed to the center of mass and the magnitudes of these accelerations are proportional to the distances of the bodies from the center of mass. This conclusion means, in particular, that the system of three bodies in the equilateral configuration can rotate as a whole (as a solid) about the center of mass under the forces of mutual gravitation. We can find the angular velocity $\omega$ of this rotation with the help of Newton's second law. Equating the product of mass of one of the bodies (say, $m_1$) and the centripetal acceleration of its rotation about $C$ to the total force exerted on this body by the other two bodies, (11.70), we obtain:

$$m_1\omega^2\mathbf{r}_{1C} = \frac{GMm_1}{R^3}\mathbf{r}_{1C}, \quad \text{whence} \quad \omega = \sqrt{\frac{GM}{R^3}} = \sqrt{\frac{G(m_1 + m_2 + m_3)}{R^3}}. \qquad (11.72)$$

Such a uniform rotation of the system in the equilateral configuration can occur only if the initial velocities of the bodies in the center-of-mass frame are exactly perpendicular to the radius vectors of the bodies relative the center of mass and the magnitudes of the velocities exactly equal the product of $\omega$ and the distances of the bodies from the center of mass. The motion is unstable. That is, if one of the above conditions is even slightly violated, the equilateral configuration soon becomes distorted and the motion of the bodies becomes irregular.

Uniform rotation is not the only possible regular periodic motion of the system in the equilateral configuration. We can show that the total gravitational force exerted on any of the bodies by the other two, being directed toward the center of mass, is inversely proportional to the square of the distance to the center of mass. Therefore under such an effectively stationary central Newtonian force (although created by the moving bodies) the body can trace a closed elliptical Keplerian orbit.

In order to prove the property of the effective gravitational field mentioned above, let us express the distance of one of the bodies (say *1*) from the center of mass $C$ in terms of the distance $R$ between any two bodies (the side of the equilateral triangle) and the masses of the bodies. Calculating the square of $\mathbf{r}_{1C}$, (11.69), and taking into account that the magnitudes of vectors $\mathbf{r}_{12}$ and $\mathbf{r}_{13}$ equal $R$, and that the angle between them equals 60°, we find $r_{1C}^2 = R^2(m_2^2 + m_3^2 + m_2m_3)/M^2$, whence

$$R^2 = \frac{M^2}{m_2^2 + m_3^2 + m_2m_3} r_{1C}^2.$$

Substituting $R$ into (11.70), we obtain:

$$\mathbf{F}_1 = m_1 \frac{G\left(m_2^2 + m_3^2 + m_2m_3\right)^{3/2}}{M^2} \frac{\mathbf{r}_{1C}}{r_{1C}^3}. \qquad (11.73)$$

Equation (11.73) shows that the total gravitational force exerted on body *1* by the other two bodies is directed to the center of mass $C$ of the system and is inversely proportional to the square of the distance between the body and the center of mass. Under this force the body moves in a Keplerian ellipse with one focus at the center of mass. The same is true for the other two bodies. And since the accelerations of the bodies, according to (11.71), are proportional to their distances from the center of mass, all three bodies can move synchronously in homothetic ellipses with a common focus at the center of mass, thus preserving the equilateral configuration. To simulate this regular, periodic motion, we should also choose certain initial velocities of the bodies. In the center-of-mass reference frame, the velocities must be proportional to the distances of the bodies from the center of mass and must make equal angles with the corresponding radius vectors. An example of a simulation of such a motion is shown in figure 9.16.

## 11.12 The sphere of gravitational action

To determine the boundaries of the region of gravitational action of a smaller body with respect to a heavier body, we consider a satellite or a spacecraft moving freely under the gravitational attraction of the Sun $S$ and a planet $P$ (figure 11.16). The mass of the spacecraft is negligible compared to the masses of the planet and Sun. Hence the Sun and the planet move as in the two-body problem, in accordance with Kepler's laws, along homothetic ellipses with a common focus located at the center of mass of the system. In other words, we deal here with a *restricted three-body problem.*

In the neighborhood of the planet it is convenient to regard the motion of the satellite as planetocentric, using the reference frame associated with the planet. The acceleration of the satellite **a** in this reference frame is produced mainly by the gravitational attraction toward the planet. According to the Newton's law of universal gravitation, this acceleration can be written in the following form:

$$\mathbf{a} = -Gm\frac{\mathbf{r}}{r^3}. \tag{11.74}$$

Here $m$ is the mass of the planet and $\mathbf{r}$ is the radius vector from the center of the planet to the satellite. An additional (perturbational) acceleration $\mathbf{a}'$ of the satellite due to the Sun in the planetocentric motion is the vectorial difference between the accelerations which the gravitational field of the Sun gives to the satellite and to the planet:

$$\mathbf{a}' = -GM\left(\frac{\mathbf{r}_s}{r_s^3} - \frac{\mathbf{R}}{R^3}\right). \tag{11.75}$$

Here $M$ is the mass of the Sun, $\mathbf{r}_s$ is the radius vector from the Sun to the satellite and $\mathbf{R}$ is the radius vector from the Sun to the planet.

Outside the sphere of gravitational action of the planet, it is more convenient to consider the motion of the spacecraft as heliocentric, using the reference frame

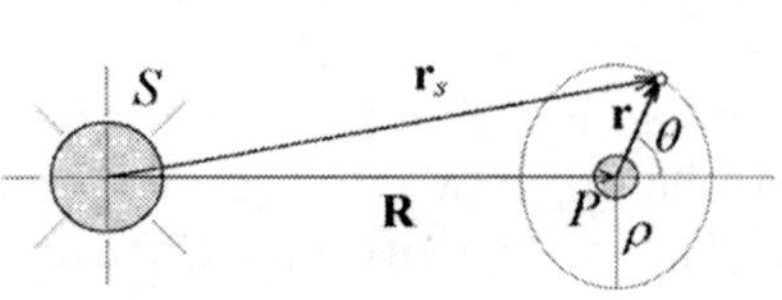


**Figure 11.16.** The sphere of gravitational action of the planet.

associated with the Sun. Here the main acceleration $\mathbf{a}_s$ of the spacecraft is produced by its gravitational attraction to the Sun:

$$\mathbf{a}_s = -GM\frac{\mathbf{r}_s}{r_s^3}. \tag{11.76}$$

In this case, the perturbational acceleration $\mathbf{a}'_s$ is the difference between the accelerations which the gravitational field of the planet gives to the spacecraft and to the Sun:

$$\mathbf{a}'_s = -Gm\left(\frac{\mathbf{r}}{r^3} - \frac{\mathbf{R}}{R^3}\right) \approx -Gm\frac{\mathbf{r}}{r^3}. \tag{11.77}$$

Assuming the mass $m$ of the planet to be small compared to the mass $M$ of the Sun, we can suppose that the radius of the planet's sphere of gravitational action with respect to the Sun is much smaller than the distance $R$ between them. This means that the values of $r$ in (11.77), which are of interest (which correspond to the boundaries of the required sphere of gravitational action of the planet), are small compared to $R$. Therefore we can neglect the second term in the parentheses, whose physical sense is the acceleration of the Sun produced by the planet. In other words, when the spacecraft is not very far from the planet, its acceleration produced by the gravitational field of the planet is much greater than the acceleration of the Sun produced by the planet.

We can regard the ratio of the magnitude of the perturbational acceleration, (11.77), to the magnitude of the main acceleration, (11.76), as a measure of the nearness of the actual motion of the spacecraft to its unperturbed Keplerian heliocentric motion. Instead of the ratio of the absolute values of these accelerations $\mathbf{a}'_s$ and $\mathbf{a}_s$, let us consider the ratio of their squares:

$$\left(\frac{a'_s}{a_s}\right)^2 = \left(\frac{m}{M}\right)^2 \frac{r_s^4}{r^4} \approx \left(\frac{m}{M}\right)^2 \frac{R^4}{r^4}. \tag{11.78}$$

Similarly, the nearness of the actual motion of the satellite to its unperturbed Keplerian planetocentric motion can be measured by the ratio of the squares of the corresponding perturbational ($\mathbf{a}'$) and main ($\mathbf{a}$) accelerations, given by (11.75) and (11.74). Before forming this ratio, we rewrite (11.75) for the case in which $m \ll M$. We represent the radius vector $\mathbf{r}_s$ in (11.75) as the sum $\mathbf{R} + r$ and in calculating the square of $\mathbf{r}_s$ we take into account that $r \ll R$ in the region in which we are interested:

$$r_s^2 = (\mathbf{R} + \mathbf{r})^2 = R^2 + 2(\mathbf{R} \cdot \mathbf{r}) + r^2 \approx R^2\left(1 + 2\frac{(\mathbf{R} \cdot \mathbf{r})}{R^2}\right),$$

whence

$$\frac{1}{r_s^3} \approx \frac{1}{R^3}\left(1 - 3\frac{(\mathbf{R} \cdot \mathbf{r})}{R^2}\right).$$

Now (11.76) can be written as follows:

$$\mathbf{a}' \approx -\frac{GM}{R^3}\left[(\mathbf{R}+\mathbf{r})\left(1 - 3\frac{(\mathbf{R}\cdot\mathbf{r})}{R^2}\right) - \mathbf{R}\right]$$
$$\approx -\frac{GM}{R^3}\left[\mathbf{r} - 3\mathbf{R}\frac{(\mathbf{R}\cdot\mathbf{r})}{R^2}\right]. \qquad (11.79)$$

The perturbational acceleration $\mathbf{a}'$, (11.79), in the motion that is described in the non-inertial planetocentric reference frame can be treated as an acceleration of the body produced by the tidal force $\mathbf{F}_{tid}$, see (11.87) in section 11.13.

We next form the ratio of the squares of the perturbational acceleration and main acceleration in the planetocentric motion:

$$\left(\frac{a'}{a}\right)^2 = \left(\frac{M}{m}\right)^2 \frac{r^4}{R^6}\left[r^2 + 3\frac{(\mathbf{R}\cdot\mathbf{r})^2}{R^2}\right] = \left(\frac{M}{m}\right)^2 \frac{r^6}{R^6}(1 + 3\cos^2\theta). \qquad (11.80)$$

Here, we have introduced the angle $\theta$ between the directions of vectors $\mathbf{R}$ and $\mathbf{r}$ (see figure 11.16) and have written their scalar product $(\mathbf{R}\cdot\mathbf{r})$ as $Rr\cos\theta$.

The boundary of the sphere of gravitational action of a planet relative to the Sun is assumed by convention to be the surface on which the ratios of the perturbational and main accelerations for the planetocentric and heliocentric motions are equal. Equating the right sides of (11.80) and (11.78), we find:

$$\frac{r}{R} = \left(\frac{m}{M}\right)^{2/5} \frac{1}{\left(1 + 3\cos^2\theta\right)^{1/10}}. \qquad (11.81)$$

For $\theta = \pm\pi/2$ (the transverse direction) the value of $r$ in (11.81) is a maximum:

$$r_{\max} = R\left(\frac{m}{M}\right)^{2/5}. \qquad (11.82)$$

This value is usually assumed as the radius of the sphere of gravitational action of a small body (a planet) relative to the massive body (the Sun).

For $\theta = 0$ (the radial direction) the radius $r$ in (11.81) is a minimum:

$$r_{\min} = R\left(\frac{m}{M}\right)^{2/5} \frac{1}{\sqrt[5]{2}}. \qquad (11.83)$$

The ratio of the maximal (transverse) radius to the minimal radius (oriented along the line joining the planet with the Sun) is only $\sqrt[5]{2} \approx 1.15$. Therefore we can assume with good precision a sphere of the radius $r_{\max}$, given by (11.82), to be the boundary of the weakly perturbed planetocentric motion.

## 11.13 The oceanic tides

The perturbations of the orbits of the Earth's satellites (in the geocentric frame of reference) caused by their gravitational attraction to the Sun and to the Moon are caused by the non-uniform character of this gravitational field on the extent of the orbit. The greater these perturbations, the more extended the satellite's orbit. On the surface of the Earth, the same physical reason is responsible for the phenomenon of oceanic tides.

The tides are manifested by alternating vertical displacements of the surface of the sea coupled with horizontal movements of the water called tidal currents. The tides are caused by the varying gravitational forces that the Moon and Sun exert on both the Earth and its oceans. More exactly, the origin of tidal phenomena is related to the inhomogeneity (non-uniformity) of the lunar and solar gravitational fields across the globe.

The gravitational force which the Moon exerts on an Earth-orbiting satellite or on any body on the surface of the Earth is much smaller than the gravitational force of the Sun. However, because the Moon is much closer to the Earth than the Sun, the inhomogeneity of the lunar gravitational field across the Earth (or across the orbit of a satellite) is considerably greater than that of the solar field. As a result, Moon-induced tides (as well as perturbations of a satellite's orbit) are more than twice as great as Sun-induced tides. Nevertheless, to arrive more easily at an understanding of the physical origin of tide-generating forces, we begin our analysis with Sun-induced tides. These are somewhat simpler to explain because the center of mass of the Sun–Earth system very nearly coincides with the center of the Sun.

We next divide the problem of tides into two parts: first we discuss the origin and properties of tide-generating forces, after which we investigate qualitatively the much more complicated case of the dynamical effect that these time-varying forces have on the ocean. For this second part, we restrict our analysis to a simplified model of the ocean as a uniform water shell of equal depth wholly covering the globe. We note that much of the confusion in the literature is related to the first (rather simple) part of this problem, which can be completely and unambiguously solved using Newtonian mechanics.

We emphasize that our treatment here is intended only to clarify the physical background of this natural phenomenon and does not describe the complete picture. The purely theoretical quantitative description of tides for a given location on the Earth, derived solely from first principles, is hardly possible because of the extremely complex structure of the oceans, the actual system that responds with tides and tidal currents to the well known tide-generating forces.

Next we discuss qualitatively the physical nature of the Sun- and Moon-induced tide-generating forces in a non-rotating geocentric frame of reference, deriving the mathematical expressions for these forces at an arbitrary point on the Earth. The static (equilibrium) distortion of the ocean surface under these forces is determined. Then we show that the same expressions for the tidal forces are applicable on the rotating Earth, and we discuss how these forces depend on time by virtue of the daily axial rotation of the Earth. We show that a uniform rotation of the system of tidal forces coupled with the apparent motion of the Sun (or Moon) can be represented as a superposition of two oscillating quadrupole systems of forces whose axes make an angle of 45° with respect to one another. Each of these systems of forces generates a steady-state forced oscillation of the ocean (a standing wave). We can treat the tidal wave circulating around the globe as a superposition of these standing waves. Finally the real-world complications of this simplified picture are discussed briefly, as well as the role of tidal friction in the evolution of the axial rotations and orbital revolutions of celestial bodies.

### 11.13.1 The origin of tidal forces: an elementary approach

The Earth as a whole moves with acceleration relative to the heliocentric inertial reference frame. This acceleration is produced by the gravitational attraction of the Earth to the Sun (and also to the Moon and to all other celestial bodies). Although the Earth travels in an almost circular orbit, its centripetal acceleration $a_0$ in this orbital motion is generated by the gravitational pull of the Sun and hence is just the acceleration of free-fall, which is independent of the orbital velocity. The Earth would move with the same acceleration were it freely falling in the gravitational field of the Sun. What is important in this problem is the acceleration, not the orbital velocity, of the Earth.

To better understand the tide-generating forces, we first use a non-rotating geocentric reference frame. Although the origin of this frame moves approximately in a circle around the Sun (more exactly, around the center of mass of the Sun–Earth system), the frame itself does not rotate because the directions of its axes are fixed relative to the distant stars. That is, the motion of this frame—revolution without rotation—is a translational (though nearly circular) motion. It reminds us of 'the circular motion of the frying pan' in the hands of a cook. With respect to the inertial space, all points of this reference frame move with an acceleration $\mathbf{a}_0$ whose magnitude and direction are the same for all the points. Any body of mass $m$ whose motion is referred to this non-inertial geocentric frame (for example, an Earth-orbiting satellite, or a drop of water in the ocean) is subject to the pseudo-force of inertia, $\mathbf{F}_{\text{in}} = -m\mathbf{a}_0$, which is independent of the position of the body relative to the Earth.

If the body were placed at the center of the Earth, this pseudo-force of inertia would exactly balance (cancel) the gravitational attraction of the body to the Sun. In other words, if we consider the Earth as a giant spaceship orbiting the Sun, a body placed at the center of this ship would seem to be weightless with respect to the gravitation of the Sun, just as astronauts on an orbital station seem to be weightless in the gravitational field of the Earth. In this reference frame, the unperturbed path of an Earth-orbiting satellite is a planar elliptical or circular closed orbit whose dimensions and orientation in space are constant—the orbit does not change over the course of time.

The pseudo-force of inertia $\mathbf{F}_{\text{in}} = -m\mathbf{a}_0$ experienced by a body in the geocentric frame (or in the frame that revolves without axial rotation about the Sun–Earth center of mass) has the same magnitude and direction everywhere on the Earth—it is independent of the coordinates. On the other hand, the gravitational pull of the Sun $\mathbf{F}_{\text{gr}}$ experienced by the body diminishes with its distance from the Sun and is directed to the Sun, and hence both the magnitude and direction of $\mathbf{F}_{\text{gr}}$ depend on the position of the body on the Earth. Except at the center of the Earth, the balance of the forces is incomplete and their combined action can be described by a *tidal force* (or *tide-generating force*). On the surface of the Earth, this force gives rise to the ocean tides. The solar and lunar gravitational perturbations of the satellite's geocentric orbit are also produced by these differential effects of gravitation. The tidal force is caused only by the inhomogeneity (non-uniformity) of the gravitational fields of the Sun and the Moon.

In other words, the tidal force at a given position near the Earth equals the vector difference of the gravitational pull the Sun exerts on an object at this position and the gravitational pull the Sun would exert on this object were it at the center of the Earth.

We may avoid using a non-inertial reference frame if we are not inclined to introduce the concept of the pseudo-force of inertia. In doing so, we can use a somewhat different language in the subsequent derivation of the tidal force: instead of discussing the vector addition of the pull of the Sun and the corresponding pseudo-force of inertia arising from the non-inertial character of the reference frame, we can use instead an inertial frame, in which the tidal force can be found by the vector subtraction of the gravitational force of the Sun on the body at its given location with the force of the Sun on the body were it located at the center of the Earth.

Indeed, when viewing the situation on the Earth from the inertial frame of reference, we can apply the Galilean law according to which, in the same gravitational field (here the field of the Sun), all free bodies experience equal accelerations. Hence the Earth as a whole and all free bodies on the Earth, being subjected to almost the same solar gravitational field, are very nearly accelerated toward the Sun. Consequently we do not particularly notice the influence of solar gravitation on what happens on the Earth. The small differences between the acceleration of the Earth as a whole and of the Earthly bodies depend on the distances of the bodies from the center of the Earth, because these differences are caused by the non-uniformity of the solar gravitational field over the extent of the Earth.

These differential effects of gravity give rise, in particular, to solar gravitational perturbations of an Earth-orbiting geocentric orbit. The tide-generating forces slightly distort the Earth's gravitational pull which governs the satellite's motion, so that after a revolution, the satellite does not return to the same point of the geocentric reference frame. On the surface of the Earth, these same forces give rise to the tides. We emphasize again that tidal forces are caused not by the Sun's gravitational field itself, but rather by the non-uniformity of this field.

Figure 11.17 illustrates the origin and properties of the tide-generating forces produced by the Sun. The free-fall acceleration of the Earth $E$ in the gravitational field of the Sun $S$ is $a_0 = GM_{\mathrm{Sun}}/R^2$, where $M_{\mathrm{Sun}}$ is the mass of the Sun and $R$ is the

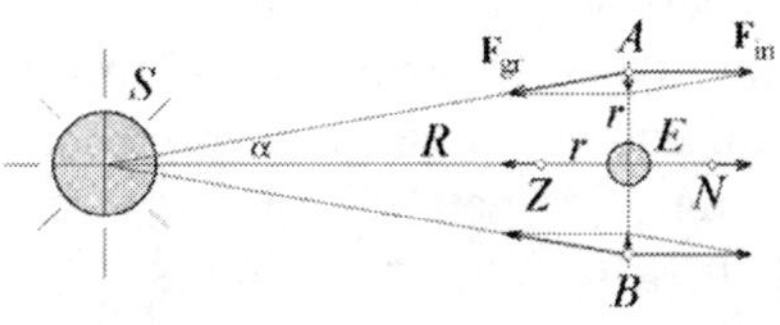


**Figure 11.17.** The tidal forces at different points over the Earth.

Sun–Earth distance. The gravitational pull of the Sun $\mathbf{F}_{gr}$ experienced by the body (e.g. by a satellite) at point $A$ almost equals the pseudo-force of inertia $\mathbf{F}_{in}$ in magnitude, because the distances of the body and the center of the Earth from the Sun are almost equal. However, the gravitational force here is not exactly opposite the inertial force. Thus their non-zero resultant, the tidal force $\mathbf{F}_A$, is directed toward the Earth. Its magnitude

$$F_A \approx ma_0\alpha = ma_0\frac{r}{R}, \tag{11.84}$$

where $\alpha = r/R$ is the angular distance between the body and the center of the Earth as seen from the Sun. The tidal force $\mathbf{F}_B$ at the opposite point $B$ equals $\mathbf{F}_A$ in magnitude and is also directed vertically downward to the Earth. On the surface of the Earth, the tidal force is directed vertically downward at all places (forming a circle) where the Sun is on the horizon at that moment.

The distance from the Sun to the body at point $Z$ (for which the Sun is at the zenith) is smaller than to the center of the Earth. Here the gravitational pull of the Sun is somewhat greater than the pseudo-force of inertia. Hence, the tidal force $\mathbf{F}_Z$ at this point is directed vertically upward (from the Earth toward the Sun). Its magnitude

$$F_Z = G\frac{mM_{Sun}}{(R-r)^2} - ma_0 = ma_0\left[\frac{R^2}{(R-r)^2} - 1\right] \approx ma_0\frac{2r}{R} \tag{11.85}$$

is approximately twice the magnitude of the tidal forces at points $A$ and $B$. Similarly, at the opposite point $N$ (for which the Sun is at nadir) the pseudo-force of inertia is greater than the gravitational pull of the Sun and so the tidal force $\mathbf{F}_N$ is also directed vertically upward from the Earth (and from the Sun). In magnitude $\mathbf{F}_N$ approximately equals $\mathbf{F}_Z$.

We note that the tidal forces are independent of the orbital motion of the Earth and would also occur if the Earth and Sun were simply falling toward each other.

The expressions for the tidal forces, $F_A = ma_0r/R = (GmM_{Sun}/R^2)(r/R)$, (11.84) and $F_Z \approx 2F_A$ given by (11.85), are also valid for the tidal forces produced on the Earth by the Moon if we replace $M_{Sun}$ by the mass of the Moon $M_{Moon}$ and $R$ by the Moon–Earth distance. There is no intrinsic difference between the origin of Sun-induced and Moon-induced tide-generating forces. In both cases, the only important factor is the acceleration of the Earth under the gravitational pull of the celestial body that causes the tides on the Earth, not the orbital velocities of both gravitationally coupled bodies (the Earth and the Sun, or the Earth and the Moon).

The tidal force experienced by any object is proportional to its distance $r$ from the center of the Earth and inversely proportional to the cube of the distance $R$ to the celestial body that causes the force, and is proportional to the mass of the source body. As noted, lunar tide-generating forces on the Earth are more than twice those of the Sun (their ratio is approximately 2.2) because the Moon is much closer to the Earth.

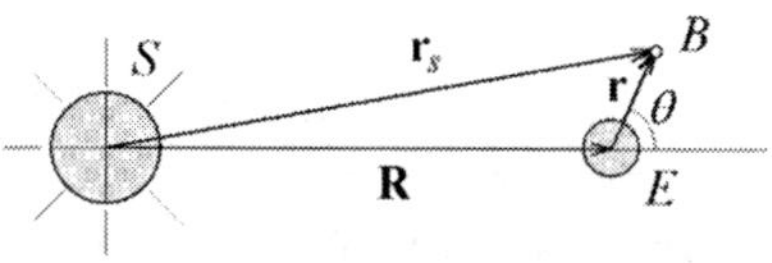


**Figure 11.18.** For the calculation of the tidal forces.

### 11.13.2 Tidal forces at an arbitrary point near the Earth

The standard derivation of tidal forces uses the tide-generating potential, for which the mathematics is somewhat simpler. However, to emphasize the physics underlying the origin of tide-generating forces, we consider the vector addition of the relevant forces.

To obtain a general mathematical expression for the tidal force at an arbitrary point $B$ over the Earth (figure 11.18), we introduce the radius vector $\mathbf{r}$ of this point relative to the center of the Earth and its radius vector $\mathbf{r}_s = \mathbf{R} + \mathbf{r}$ relative to the center of the Sun $S$, where $\mathbf{R}$ is the radius vector of the Earth relative to the Sun. The tidal force $\mathbf{F}_{\text{tid}}$ experienced by a body of mass $m$ at point $B$ (in the non-inertial geocentric frame) is the result of its gravitational attraction $\mathbf{F}_{\text{gr}}$ to the Sun and the pseudo-force of inertia $\mathbf{F}_{\text{in}} = -m\mathbf{a}_0 = -GmM_{\text{Sun}}\mathbf{R}/R^3$:

$$\mathbf{F}_{\text{tid}} = \mathbf{F}_{\text{gr}} + \mathbf{F}_{\text{in}} = -GmM_{\text{Sun}}\left(\frac{\mathbf{r}_s}{r_s^3} - \frac{\mathbf{R}}{R^3}\right). \tag{11.86}$$

Next we express $\mathbf{r}_s$ in (11.86) as the sum $\mathbf{R} + \boldsymbol{r}$ and calculate the square of $\mathbf{r}_s$ taking into account that $r \ll R$:

$$r_s^2 = (\mathbf{R} + \mathbf{r})^2 = R^2 + 2(\mathbf{R} \cdot \mathbf{r}) + r^2 \approx R^2\left(1 + 2\frac{(\mathbf{R} \cdot \mathbf{r})}{R^2}\right).$$

Therefore,

$$\frac{1}{r_s^3} \approx \frac{1}{R^3}\left(1 - 3\frac{(\mathbf{R} \cdot \mathbf{r})}{R^2}\right).$$

Substituting this expression for $1/r_s^3$ into (11.86) yields:

$$\begin{aligned} \mathbf{F}_{\text{tid}} &\approx -G\frac{mM_{\text{Sun}}}{R^3}\left[(\mathbf{R} + \mathbf{r})\left(1 - 3\frac{(\mathbf{R} \cdot \mathbf{r})}{R^2}\right) - \mathbf{R}\right] \\ &\approx -G\frac{mM_{\text{Sun}}}{R^3}\left[\mathbf{r} - 3\mathbf{R}\frac{(\mathbf{R} \cdot \mathbf{r})}{R^2}\right]. \end{aligned} \tag{11.87}$$

We note that the main contributions of $\mathbf{F}_{\text{gr}}$ and $\mathbf{F}_{\text{in}}$ to $\mathbf{F}_{\text{tid}}$, whose magnitudes are inversely proportional to $R^2$, cancel in (11.87). This cancellation corresponds to the aforementioned state of weightlessness that we experience on the spaceship 'Earth' with respect to the Sun's gravity. The magnitude of the remaining force in (11.87) is inversely proportional to $R^3$, that is, to the *cube* of the distance from the Sun (generally, from the celestial body which is the source of the tidal force).

For points $A$ and $B$ considered above (see figure 11.17), radius vector $\mathbf{r}$ is perpendicular to $\mathbf{R}$ and the scalar product $\mathbf{R} \cdot \mathbf{r}$ in (11.87) is zero. Hence, at these points the tidal force is directed opposite to $\mathbf{r}$ (that is, vertically downward) and its magnitude equals $GmM_{\text{Sun}}r/R^3$. For the points $Z$ and $N$, the tidal force, according to (11.87), is directed along $\mathbf{r}$ (that is, vertically upward) and its magnitude $2GmM_{\text{Sun}}r/R^3$ is two times greater than at points $A$ and $B$. We see that at these four points, the general result given by (11.87) agrees with the simpler calculations presented above, see (11.84) and (11.85).

### 11.13.3 The horizontal and vertical components of the tidal force

The Sun-induced tide-generating forces exerted on the Earth have a quadrupole character: they stretch the Earth along the Sun–Earth line and squeeze the Earth in the directions perpendicular to that line, as shown in figure 11.19. Because of the axial symmetry with respect to the Sun–Earth line, the vertical and horizontal components of the tidal force depend only on the angle $\theta$ shown in figure 11.18 (and on the distance $r$ from the center of the Earth). The angle $\theta$ determines the position of the mass point $m$ on or near the surface of the Earth measured from this line. Certainly, the system of the Moon-induced tidal forces also has a quadrupole character.

Multiplying the right side of (11.87) by the unit vector $\mathbf{r}/r$, we obtain the following dependence of the vertical component of the tidal force $\mathbf{F}_{\text{tid}}$ on the angle $\theta$ between $R$ and $r$:

$$(F_{\text{tid}})_{\text{vert}} = G\frac{mM_{\text{Sun}}}{R^3}\, r(3\cos^2\theta - 1) = F_{\text{gr}}\frac{r}{R}(3\cos^2\theta - 1). \qquad (11.88)$$

At all the points for which $\cos\theta = \pm 1/\sqrt{3}$, the tidal force is directed horizontally.

For the horizontal component of the tidal force at an arbitrary point $B$ whose geocentric position is determined by coordinates $r$ and $\theta$ (see figure 11.18), (11.87) yields:

$$(F_{\text{tid}})_{\text{hor}} = -3G\frac{mM_{\text{Sun}}}{R^3}\, r\cos\theta\sin\theta = -\frac{3}{2}F_{\text{gr}}\frac{r}{R}\sin(2\theta). \qquad (11.89)$$

The horizontal component $(F_{\text{tid}})_{\text{hor}}$ is maximal at all the points of the Earth for which $\theta = \pm 45°$ and $\theta = \pm 135°$. The horizontal component of the tidal force has its

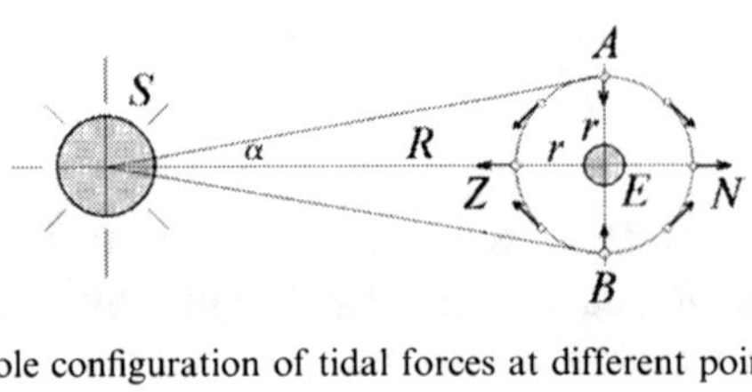


**Figure 11.19.** The quadrupole configuration of tidal forces at different points over the Earth's surface.

maximum value $(3/2)(r/R)F_{gr} = (3/2)(r/R)GmM_{Sun}/R^2$. This maximal horizontal component of the solar tide-generating force causes a deviation of the plumb line from the direction of the Earth's own gravity by only 0.008′.

The magnitude of the tidal force experienced by a satellite is proportional to its distance $r$ from the center of the Earth and inversely proportional to the *cube* of the distance $R$ to the celestial body that causes the force. The above expressions are also valid for the tidal forces produced by the Moon if we replace there the mass of the Sun $M_{Sun}$ by the mass of the Moon, and the Sun–Earth distance $R$ with the Moon–Earth distance. Although the mass of the Moon is much smaller than that of the Sun, the lunar perturbations of a satellite's orbit are more than twice as great as the solar perturbations because the Moon is much closer to the Earth. A simulation of the lunar perturbations of an Earth-orbiting satellite is described in section 8.6.

Expression (11.88) for the vertical component of the tidal force $\mathbf{F}_{tid}$ can be transformed to the following form:

$$(F_{tid})_{vert} = G\frac{mM_{Sun}}{R^3}\, r(3\cos^2\theta - 1) = \frac{3}{2}G\frac{mM_{Sun}}{R^2}\,\frac{r}{R}\left(\cos 2\theta + \frac{1}{3}\right). \tag{11.90}$$

The last term in the right-hand expression of (11.90) is independent of $\theta$ and is thus independent of time on the spinning Earth. It can therefore be dropped as far as the tides are concerned. This term in the vertical component of the tidal force is the same everywhere on the Earth (for a given value of $r$) and adds only a tiny constant value to the vertical force of the Earth's gravity $mg$ (about ten million times smaller than $mg$). Thus, the vertical and horizontal components of the tidal force exerted on a body of mass $m$ located at a position determined by angle $\theta$ and radius $r$ are given by the expression:

$$(F_{tid})_{vert} = \frac{3}{2}F_{gr}\frac{r}{R}\cos(2\theta), \qquad (F_{tid})_{hor} = -\frac{3}{2}F_{gr}\frac{r}{R}\sin(2\theta), \tag{11.91}$$

where $F_{gr} = GmM_{Sun}/R^2$ is the total gravitational pull of the Sun (or the Moon) experienced by the body anywhere on the Earth. This representation of the tide-generating force is especially convenient because (11.91) defines a tidal force vector whose magnitude $(3/2)(r/R)F_{gr} = (3/2)GmM_{Sun}/R^2(r/R)$ is independent of the angle $\theta$: the tidal forces at all points that lie at a given distance $r$ from the Earth's center are equal in magnitude and differ only in direction.

Equations (11.91) are also valid for the tidal forces produced by the Moon, provided we replace the mass of the Sun $M_{Sun}$ by the mass of the Moon $M_{Moon}$ and the Sun–Earth distance $R$ by the Moon–Earth distance. In this case the angle $\theta$ in (11.91) determines the position of the body relative to the Moon–Earth line.

The tide-generating force of the Moon, $F_{tid} = (3/2)GmM_{Moon}r_0/R^3$, experienced by a body of mass $m$ on the surface of the Earth (here $r_0$ is the Earth's radius) is very small compared to its weight, that is, with the Earth's force of gravity

$F_{\rm grav} = mg = GmM_{\rm Earth}/r_0^2$. If we let the ratio $M_{\rm Moon}/M_{\rm Earth} = 1/81$ and the mean Earth–Moon distance $R = 60\, r_0$ (actually this distance varies between $57\, r_0$ and $63.7\, r_0$ because of the elliptical shape of the Moon's orbit), we obtain:

$$\frac{F_{\rm tid}}{F_{\rm grav}} = \frac{F_{\rm tid}}{mg} = \frac{3}{2}\frac{M_{\rm Moon}}{M_{\rm Earth}}\frac{r_0^3}{R^3} = 8.6 \cdot 10^{-8}. \tag{11.92}$$

Although the tide-generating forces are very small in comparison with the Earth's force of gravity (on the surface of the Earth the lunar tidal force at its maximum being only about $10^{-7}$ times the Earth's gravitational force), their effects on the water of the ocean are considerable because of their horizontal component, which is orthogonal to the Earth's gravitational field and varies with time periodically because of the Earth's axial rotation. This horizontal component shifts the ocean water around the globe. We emphasize again that the horizontal (tangential to the surface) components of the tidal forces are much more influential on the ocean tides and on the orbits of Earth satellites compared to the vertical (radial) components of the tidal forces, which only modify the Earth's gravitational force slightly.

### 11.13.4 The static distortion of the water surface

To estimate the static (equilibrium) distortion of the ocean's surface due to the tidal forces, we can use the hypothetical situation of a non-rotating planet on which the tide-generating forces are nearly time-independent. From the symmetry of the tidal forces, (11.91), we can assume that the distorted surface has an ellipsoidal shape given by the expression

$$r(\theta) = r_0 + a\cos 2\theta, \tag{11.93}$$

where $2a \ll r_0$ is the difference in the static maximal and minimal levels at points $Z$ and $A$ (see figure 11.19). Hence we can write for the small inclination $\alpha$ of the water surface with respect to the horizon:

$$\alpha = \frac{1}{r}\frac{{\rm d}r(\theta)}{{\rm d}\theta} \approx -\frac{2a}{r_0}\sin 2\theta. \tag{11.94}$$

We see that the water surface is horizontal ($\alpha = 0$) at $\theta = 0$ and $\theta = 90°$ (points $Z$ and $A$). The angle $\alpha$ is maximum and equals $2a/r_0$ at $\theta = 45°$ and at $\theta = 135°$, where the tidal force is directed horizontally. In equilibrium the distorted water surface is orthogonal to the plumb line. The plumb line shows the direction of the vector sum of the Earth's gravity and the tidal force. A small departure of the plumb line from the direction of the Earth's gravity is caused by the horizontal component of the tidal force. Therefore, the angle $\alpha$ equals the ratio of the horizontal tidal force $(F_{\rm tid})_{\rm hor}$ to the force of the Earth's gravity $F_{\rm grav} = mg$. If we equate $\alpha = 2a/r_0$ at $\theta = 45°$ to $(F_{\rm tid})_{\rm hor}/F_{\rm grav}$ and take into account that for Sun-induced tides,

$(F_{\rm tid})_{\rm hor}/F_{\rm grav} = (3/2)\ (M_{\rm Sun}/M_{\rm Earth})(r_0^3/R^3)$, we find for the maximal static level difference $2a$ at points $Z$ and $A$:

$$2a = \frac{3}{2} r_0 \frac{M_{\rm Sun}}{M_{\rm Earth}} \frac{r_0^3}{R^3}. \qquad (11.95)$$

Equation (11.95) yields $2a = 0.24$ m. A similar expression is valid for the static distortion of the ocean surface due to the lunar tidal force, and yields $2a = 0.54$ m for the Moon-induced static distortion. In section 11.13.6, the equation for this static distortion is also derived from the tide-generating potential.

### 11.13.5 Tidal forces on the rotating Earth

In the above treatment of tide-generating forces we have used a revolving but non-rotating geocentric reference frame. The origin of this frame moves in a circle around the Sun–Earth (Moon–Earth) center of mass, but the frame itself does not rotate because the directions of its axes are fixed relative to the distant stars. That is, the frame moves translationally in a circle. This reference frame is convenient for the analysis of the motion of an artificial satellite. If we ignore the perturbations caused by tidal forces, an Earth satellite traces out a closed elliptical orbit relative to this reference frame.

To introduce tidal forces on the rotating Earth, we must use a true geocentric frame of reference that plays a part in the daily rotation of the Earth. This frame is non-inertial and hence we should be concerned with the acceleration of its different points. We can consider the motion of the Earth (and of the geocentric reference frame) as consisting of two components. The first is the component considered above, namely translational motion (revolution without rotation) about the Sun–Earth (Moon–Earth) center of mass. The second component is a uniform daily rotation (spin) of the Earth about an axis passing through the center of the Earth.

Both these motions of the Earth are important in the problem of tides, but the roles they play are quite different. The acceleration $\mathbf{a}_0$ related to the translational motion is responsible for the origin of the uniform pseudo-force of inertia $\mathbf{F}_{\rm in} = -m\mathbf{a}_0$, whose action on a body on the Earth, combined with the non-uniform gravitational pull of the Sun (Moon), is described by the tidal force $\mathbf{F}_{\rm tid}$ considered previously. We note again that only the acceleration $\mathbf{a}_0$ of this translational motion is important, not the orbital velocity of the Earth.

To avoid the confusion often encountered in the literature, we must be careful with our definitions. In discussing tides, we should be concerned only with those gravitational and inertial forces that depend on the apparent position of the celestial body that produces the tide. The axial rotation of the Earth is related to centripetal acceleration and gives rise to centrifugal pseudo-forces that increase in proportion to the distance from the Earth's axis. The centrifugal force of the Earth's daily rotation is generally much greater in magnitude than the tidal forces. Because of the centrifugal forces, the equilibrium shape of the Earth differs slightly from an ideal sphere—it is approximately an ellipsoid of rotation whose equatorial diameter is a bit greater than the polar diameter. The centrifugal effect of the Earth's daily rotation causes an equatorial bulge, which is the principal departure of the Earth from its spherical shape.

But we are not concerned here with this constant distortion of the Earth because this distortion is independent of the apparent position of the celestial body that produces the tides. Therefore the centripetal acceleration of the axial rotation adds nothing to tidal forces. However, the daily rotation of the Earth makes tidal forces time dependent because the pattern of tidal forces on the Earth is coupled to the apparent positions of the Sun and Moon. A dynamical response of the oceanic waters on the spinning Earth to these time dependent forces is the essence of the phenomenon of tides.

Thus, in the problem of tides, expressions for the tide-generating forces $(F_{\text{tid}})_{\text{hor}}$ and $(F_{\text{tid}})_{\text{vert}}$ in (11.91) are also applicable to the true geocentric frame of reference, which plays a part in the daily axial rotation of the Earth. The system of tidal forces shown in figure 11.19 being coupled to the apparent position of the Sun (Moon), rotates rigidly together with the Earth–Sun (Earth–Moon) line. For simplicity, we shall consider the case in which the celestial body—the source of tidal forces (the Sun or Moon)—occurs in the equatorial plane of the Earth. Although the system of tidal forces rotates as a whole with the angular velocity $\Omega$ of the Earth's axial rotation, that is, with a period of $2\pi/\Omega$, the true period $T$ of the variation of the tidal forces on the Earth equals half this value ($T = \pi/\Omega$) because of the quadrupole symmetry of the system of forces (the semidiurnal tide). For the Sun-induced tidal forces the period equals 12 h. For the Moon-induced tidal forces the period is 12 h 25 min—the difference between the two periods is due to the orbital motion of the Moon.

If we fix a point on the equator of the Earth, the local tidal force vector executes a uniform rotation in the vertical plane, making two complete revolutions during a day. The simulation available online[2] clearly shows how the daily rotation of the whole system of tidal forces produces this doubly fast uniform rotation of the tidal force at a given equatorial point, as seen by an observer on the spinning Earth. Because of this periodic dependence on the time, the tidal forces, in spite of their small magnitude compared even to the centrifugal force of inertia, produce the oceanic tides.

To find analytical expressions for the time dependence of the tidal forces at a given point in the equatorial plane of the spinning Earth, we substitute $\theta(t) = \Omega t$ in (11.91) for $(F_{\text{tid}})_{\text{hor}}$ and $(F_{\text{tid}})_{\text{vert}}$. This substitution yields the following expressions for the point of the equator at which the Sun culminates (passes through its zenith) at $t = 0$:

$$(F_{\text{tid}})_{\text{vert}}(t) = Ar\cos(2\Omega t), \qquad (F_{\text{tid}})_{\text{hor}}(t) = -Ar\sin(2\Omega t), \tag{11.96}$$

where $A = (3/2)F_{\text{gr}}/R = (3/2)GmM_{\text{Sun}}/R^3$. At any other equatorial point of the Earth, the tidal force vector also rotates in the vertical plane with angular velocity $2\Omega$.

[2] See on the web http://faculty.ifmo.ru/butikov/Projects/Tides0.html

That is, all the vectors of tidal forces at different points rotate synchronously but with different phases.

### 11.13.6 The potential function for tidal forces

An approach often used in the literature for deriving an expression of the tidal force is to begin with the potential energy of a body under the influence of tide-generating forces. This approach is simpler than that presented above. However, we have chosen the above approach because it does not obscure the underlying physics and consequently may be considered advantageous for developing intuition. Nevertheless, for completeness, we introduce here the potential function $U_{\text{tides}}(r,\theta)$ and show how it can be used in calculating the equilibrium shape of the surface of the ocean and the static distortion of the water under tidal forces.

The components of the force that lie in the equatorial plane are given in (11.91). We may consider these components as the negative gradients of the corresponding potential function $U_{\text{tides}}(r,\theta)$:

$$\left(F_{\text{tid}}\right)_{\text{vert}}(\theta) = Ar\cos(2\theta) = -\frac{\partial U_{\text{tides}}(r,\theta)}{\partial r},$$
$$\left(F_{\text{tid}}\right)_{\text{hor}}(\theta) = -Ar\sin(2\theta) = -\frac{1}{r}\frac{\partial U_{\text{tides}}(r,\theta)}{\partial\theta}. \tag{11.97}$$

Therefore, the desired potential function for the tidal forces can be written as:

$$U_{\text{tides}}(r,\theta) = -\frac{1}{2}Ar^2\cos 2\theta = -\frac{3}{4}\frac{GmM_s}{R^3}r^2\cos 2\theta. \tag{11.98}$$

The restoring forces that limit the tidal distortion of the water's surface are due to the Earth's gravity. In the absence of tidal forces, the shape of the oceanic water surface under the Earth's gravity would be an ideal sphere. If the Earth were not rotating relative to the Earth–Sun line, the static distortion of the water surface covering the globe would be the surface of equal total potential:

$$U(r,\theta) = U_0(r) + U_{\text{tides}}(r,\theta) = \text{const}, \tag{11.99}$$

where $U_0(r) = mgr$ is the spherically symmetric potential function of the Earth's gravity which yields the radial component of the Earth's gravitational force $\mathrm{d}U_0(r)/\mathrm{d}r = mg$. Thus,

$$U(r,\theta) = mgr - \frac{1}{2}Ar^2\cos 2\theta. \tag{11.100}$$

In particular, at points $Z$ and $A$ (see figure 11.19) of the water surface, the values of the total potential function, (11.100), are equal: $U(r_Z,\pi) = U(r_A,\pi/2)$, from whence we obtain

$$mgr_Z - \frac{1}{2}Ar_Z^2 = mgr_A + \frac{1}{2}Ar_A^2,$$
$$mgr_Z - mgr_A = \frac{1}{2}A(r_Z^2 + r_A^2). \tag{11.101}$$

We can use this condition, (11.101), to determine the static equilibrium distortion under the tidal forces of the otherwise spherical ocean surface. Let the radii of the distorted water surface at points $Z$ and $A$ be $r_Z = r_0 + a$ and $r_A = r_0 - a$, respectively, where $r_0$ is the radius of the undistorted surface. Then $2a$ is the static level difference at points $Z$ and $A$ in which the level is maximum and minimum, respectively. Thus, from (11.101) we have $2mga = (1/2)A(r_Z^2 + r_A^2) \approx Ar_0^2$, and for $2a$ we obtain:

$$2a = \frac{Ar_0^2}{mg} = \frac{3}{2} r_0 \frac{F_{\text{Sun}}}{mg} \frac{r_0}{R}. \tag{11.102}$$

We note that $F_{\text{Sun}}/mg = (M_{\text{Sun}}/M_{\text{Earth}})(r_0^2/R^2)$, so the static distortion of the ocean surface under the Sun-induced tidal forces can also be expressed as:

$$2a = \frac{3}{2} r_0 \frac{M_{\text{Sun}}}{M_{\text{Earth}}} \frac{r_0^3}{R^3}. \tag{11.103}$$

This expression is the same as (11.95) derived by requiring that in equilibrium, the surface of the ocean be orthogonal to the vector sum of the Earth's gravitational force and the tidal force.

### 11.13.7 The natural wave and the driving tidal forces

Most textbook authors oversimplify the problem of tides and consider (after Newton and Bernoulli) only the so-called static (or equilibrium) theory of tides, which treats the ocean surface as a liquid ellipsoid stretched along the Earth–Moon (Earth–Sun) line, as if this surface were always in equilibrium under the Earth's force of gravity and the tidal forces produced by the Moon (Sun). In this approach, the tidal bulges are aligned with the Earth–Moon (or Earth–Sun) axis. Therefore on the spinning Earth the moments of high water at a given location should coincide with the upper and lower culminations of the Moon (Sun), that is, when the Moon (Sun) passes through its zenith and nadir. Observations do not agree with this prediction. Instead, almost the opposite is usually observed: the moments of low tide occur approximately at the culminations of the Moon.

A complete theory of the tides should take into account the dynamical response of the ocean to the time dependent generating forces. The dynamical theory of tides (first suggested by Laplace and developed by Airy) treats the tides as a steady-state forced motion (under varying tidal forces) of a dynamical system (the ocean). Such a theory predicts a resonant growth of the steady-state amplitude in cases when the driving period approaches the period of natural oscillations.

To avoid the complications related to the three-dimensional character of the problem and to explain the physical aspect of the dynamical theory using the simplest possible model, we imagine, following Airy, water in a wide canal of uniform depth encircling the entire Earth along the equator. Imagine the water surface in this canal being distorted statically under the tide-generating forces so that two bulges form on opposites sides of the Earth, changing the shape of the water surface from circular to elliptical. If the forces maintaining this shape suddenly vanish, the Earth's gravity would make the distorted surface restore its equilibrium, circular shape. The water would start to flow and the bulges disappear so that after a time, namely a quarter period, the water surface would become circular. But because the water continues to move, after another quarter period the bulges reappear in new positions showing an elliptical distortion of the surface along the line perpendicular to the line of the original distortion. Then the motion repeats itself in reverse. This motion of water in the circular canal is a gravitational standing surface wave whose wavelength equals a half-circumference of the globe. Such a mode of oscillation is characterized by a certain natural period.

The superposition of two such standing waves whose phases differ by $\pi/2$ and whose elliptical axes are separated by 45° produces a circulating (traveling) wave of constant elliptical shape and a wavelength equal to half of the Earth's circumference. The two opposite bulges in the water surface travel with this wave around the globe preserving their height and shape.

An essential point in explaining the steady-state phase shift between the moments of high tide and the culmination of the Moon (Sun) is the relation between the natural period $T_0$ of this circulating wave and the period $T$ of the tide-generating driving forces. It is possible to estimate $T_0$ as the time taken by the circulating surface wave to travel along half the globe. In the limiting case of very long waves on the surface of shallow water ($\lambda \gg h$), the speed of the wave is determined by the Earth's gravity $g$ and depth $h$, and is independent of $\lambda$. From hydrodynamics we know that this speed equals $\sqrt{gh}$. We assume that the mean value $h$ of the ocean depth is 3.5 km. During a period $T_0$ the wave travels half the circumference of the globe $\pi r_0$, and hence $T_0 = \pi r_0/\sqrt{gh} \approx 30$ h. Thus, the approximately 12 h driving external period $T$ is less than the natural period $T_0$ of the water free oscillation in the equatorial canal surrounding the Earth.

We emphasize that it is the shape of the surface (the wave) that circulates around the globe, not the water itself. Relative to the Earth, points on the surface of the ocean (water particles) execute oscillatory motions in closed paths that are considerably stretched horizontally. These back and forth motions constitute the tidal currents. On average, the water is stationary in the geocentric frame.

To obtain a dynamical picture of tides on the rotating Earth, we should use the reference frame that rotates with the Earth. Relative to this frame, the quadrupole system of tide-generating forces, being coupled to the position of the Sun (Moon), rotates as a whole while the Sun (Moon) travels along its apparent daily path around the Earth. This rotation of the forces occurs at an angular velocity $\Omega$, the angular velocity of the Earth's daily rotation (or the difference between $\Omega$ and the angular velocity of the Moon in its orbit for Moon-induced tides). Such a uniform rigid rotation of the system of mutually fixed vectors can be represented as a superposition of two oscillating quadrupole systems of forces (with a frequency $\omega = 2\Omega$) that do not rotate and whose axes make an angle of 45° to one another. At each point one of these forces oscillates along the radial (vertical) direction, while the other force oscillates along the tangential (horizontal) direction. The oscillations of these orthogonal components occur a quarter period out of phase. At any given point in the equatorial plane, the vector sum of these mutually orthogonal oscillating forces produces a force of constant magnitude whose direction rotates uniformly following the apparent motion of the Sun (Moon), but with angular velocity $\omega = 2\Omega$. For different points on the Earth, the phases of these rotating vectors differ. This behavior of the tidal forces can clearly be observed with the help of the above mentioned simulation program developed by the author and available online at http://faculty.ifmo.ru/butikov/Projects/Tides0.html.

### 11.13.8 The tides as forced oscillations of the ocean

What is really of interest is the steady-state forced oscillation of the ocean surface due to the time dependent tidal forces. Each of the two oscillating systems of forces described above excites a mode of forced oscillation of the water in the equatorial canal, specifically the mode of the same symmetry as is characteristic of the corresponding system of driving forces. These modes have elliptical shapes, much like the natural oscillations considered above, namely, the elliptical standing waves whose axes make an angle of 45° with one another. Nevertheless, we can consider these modes to be orthogonal in the sense that their spatial forms are described by eigenfunctions forming an orthogonal basis in the function space. The two forced oscillations in this linear system, each excited by one system of oscillating driving tidal forces, are independent of one another and the resulting forced motion is a superposition of these forced oscillations.

Any steady-state forced oscillation occurs exactly with the period of the driving force. The amplitude and phase lag of the oscillation depend on the amplitude of the driving force, on the damping factor and, more importantly, on the relation between the driving and natural periods. The two systems of oscillating driving tidal forces are characterized by equal amplitudes and frequencies. Also the natural frequencies and damping factors of both excited modes are equal. Hence both excited modes also have equal amplitudes and equal phase delays behind the corresponding driving forces. The superposition of these modes produces a forced circulating (traveling) elliptical wave that has the same phase relation with the rotating driving forces as is characteristic of forced oscillations in general.

If we ignore friction (dissipation of mechanical energy in the excited wave motion), the forced motion occurs exactly in phase with the driving force, provided the driving period is longer than the natural period. Otherwise the forced motion occurs in the opposite phase with respect to the driving force. For the simplified model of tides in the equatorial canal of uniform depth (and also for an Earth covered everywhere by an ocean of uniform depth), the natural period of free oscillation is longer than the 12 h driving period. Thus the dynamical theory predicts in this case a stationary circulating elliptically shaped wave whose axis (the line of tidal bulges) is perpendicular to the Earth–Sun (Earth–Moon) line.

On the other hand, the natural period of an elastic wave in the crust of the Earth is shorter than the 12 h period of the tidal forces. Hence, in the frictionless model, bulges in the Earth's crust are oriented along the Earth–Sun (Earth–Moon) line. Observations show that the solid body of the Earth actually experiences twice-daily tides with a maximum amplitude of about 30 cm whose bulges lag approximately 3° behind the Earth–Moon line.

### 11.13.9 A mathematical description of the forced oscillations

Each of the partial forced oscillations can be described by a differential equation of a linear oscillator. Let $q_1(t)$ be the normal coordinate describing the first forced oscillation whose elliptical shape is characterized by a major axis oriented along the Earth–Sun line (and in the perpendicular direction after a half period) and let $q_2(t)$ be the normal coordinate describing the second oscillation with the axis inclined through 45° with respect to the Earth–Sun line. A disturbance of the water surface caused by the first oscillation can be described by $\Delta r_1(\theta, t) = q_1(t)\cos(2\theta)$, which gives the small vertical displacement of the surface at an arbitrary point $(r_0, \theta)$ of the equator. Similarly, the second oscillation causes a distortion of the surface described by $\Delta r_2(\theta, t) = q_2(t)\cos(2\theta)$. The forced oscillations experienced by the normal coordinates $q_1(t)$ and $q_2(t)$ are periodic (steady-state) partial solutions of the two differential equations:

$$\begin{aligned} \ddot{q}_1 + 2\gamma\dot{q}_1 + \omega_0^2 q_1 &= \omega_0^2 a \cos \omega t, \\ \ddot{q}_2 + 2\gamma\dot{q}_2 + \omega_0^2 q_2 &= \omega_0^2 a \sin \omega t. \end{aligned} \tag{11.104}$$

Here $\omega_0$ is the natural frequency of the corresponding mode ($\omega_0 = 2\pi/T_0 = 2\sqrt{gh}/r_0$), $\gamma$ is the damping constant, $\omega = 2\Omega$ is the driving frequency and $a$ is the magnitude of the equilibrium distortion of the ocean surface under the static system of tidal forces (that is, the distortion for the planet whose axial rotation is synchronized with its orbital revolution). The theoretical value of $a$ is given by (11.102) or (11.103). Although the values of $a$ and of $\omega = 2\Omega$ (twice the angular velocity $\Omega$ of the Earth's daily rotation) are fairly well known, the situation is quite different regarding the values of $\omega_0$ and $\gamma$.

In the hypothetical limiting case of extremely slow rotation of the Earth, the steady-state solution of (11.104) is $q_1(t) = a \cos \omega t = a \cos 2\Omega t$, $q_2(t) = a \sin \omega t = a \sin 2\Omega t$. This solution describes the quasi-static elliptical distortion whose axis follows adiabatically the slowly rotating Earth–Sun (Earth–Moon) line. The major axis of the ellipse at any moment is oriented along this line. The displacement of the water level from its mean position in the equatorial plane in this limiting case is given by:

$$\begin{aligned} \Delta r(\theta, t) &= \Delta r_1(\theta, t) + \Delta r_2(\theta, t) = q_1(t)\cos 2\theta + q_2(t)\sin 2\theta \\ &= a(\cos 2\Omega t \cos 2\theta + \sin 2\Omega t \sin 2\theta) = a \cos 2(\Omega t - \theta). \end{aligned} \tag{11.105}$$

To find the distortion of the water surface for an arbitrary value of $\omega$, we can use the relevant well known steady-state solutions to (11.104) for the normal coordinates $q_1(t)$ and $q_2(t)$:

$$q_1(t) = q_0 \cos(\omega t - \delta), \qquad q_2(t) = q_0 \sin(\omega t - \delta), \tag{11.106}$$

where their common amplitude $q_0$ and phase lag $\delta$ are given by:

$$q_0 = \frac{\omega_0^2 a}{\sqrt{\left(\omega_0^2 - \omega^2\right)^2 + 4\gamma^2\omega^2}}, \qquad \tan\delta = \frac{2\gamma\omega}{\omega_0^2 - \omega^2}. \tag{11.107}$$

Therefore the resulting distortion of the water surface under the tidal forces is given by

$$\begin{aligned}\Delta r(\theta, t) &= \Delta r_1(\theta, t) + \Delta r_2(\theta, t) = q_1(t)\cos 2\theta + q_2(t)\sin 2\theta \\ &= q_0[\cos(2\Omega t - \delta)\cos 2\theta + \sin(2\Omega t - \delta)\sin 2\theta] \\ &= q_0 \cos 2(\Omega t - \delta/2 - \theta).\end{aligned} \tag{11.108}$$

We see from (11.108) that at any time $t$ the maximum (high water) of the tidal wave circulating around the Earth is located at the position defined by the angle $\theta_{max} = \Omega t - \delta/2$. That is, the position of this maximum lags behind the Sun (Moon) by the angle $\delta/2$. If $\gamma \ll \omega$, it follows from (11.108) that this retarding angle is almost zero for the case $\omega < \omega_0$. In other words, the marine tide would be nearly the equilibrium tide with the high-water time coinciding with culminations of the Sun (Moon) if the natural period of the circulating wave were less than the 12 h driving period (that is, if $T_0 < T$). However, for our model of the ocean, we estimate the natural period to be approximately 30 h. Therefore the situation corresponds to $\omega > \omega_0$ ($T < T_0$), when the steady-state forced oscillations occur nearly in the opposite phase relative to the driving force. In this case the tide should be inverted with respect to the equilibrium one. The retarding angle $\delta/2$ approaches $\pi/2$ according to (11.109), which means that for a given equatorial point, the high water occurs when the Sun (Moon) is almost at the horizon (rather than at zenith or nadir).

At any given place on the equator, it follows from (11.108) that the water level (above the average value) varies with time $t$ according to $z(t) = q_0\cos(2\Omega t - \delta)$, where $t = 0$ corresponds to the culmination of the Sun (Moon) at the place in question. We can expect that for the model of a water canal of uniform depth, the value of $q_0$ given by (11.107) is more or less reliable because hydrodynamics allows us to estimate the natural frequency $\omega_0 = 2\pi/T_0 = 2\sqrt{gh}/r_0$ by using the known speed $v = \sqrt{gh}$ of very long gravitational waves. However, considerable uncertainty is related to the damping factor $\gamma$. If we assume that the damping is small ($\gamma \ll \omega_0$), we can conclude that the orientation of the tidal bulges deviates only slightly from the line perpendicular to the Sun–Earth (Moon–Earth) line, but the particular value of this deviation remains indefinite.

In the above discussion, we considered only the steady-state oscillation of the ocean surface (the stationary wave), assuming that the transient is already over. For this steady motion to establish itself, some friction (even if very small) is necessary. In the problem under consideration, we are concerned with the water motion caused solely by the eternally lasting tidal forces and therefore we have had centuries and even millennia to wait for the fading away of the transient. Therefore our use of the steady-state solution is appropriate for tides. We also emphasize that in the dynamical theory of tides, the driving tide-generating forces are perfectly well known, so that most uncertainties originate primarily from a very poor correspondence between the simple model of the dynamical system and the real oceans of the Earth.

### 11.13.10 Real-world complications

The pattern of tide-generating forces is coupled to the position of the Moon (and the Sun) with respect to the Earth. For any place on the Earth's surface, the relative position of the Moon has an average periodicity of 24 h 50 min. The lunar tide-generating force experienced at any location has the same periodicity. When the Moon is in the plane of the equator, the force runs through two identical cycles within this time interval because of the quadrupole symmetry of the global pattern of tidal forces. Consequently, the tidal period is 12 h 25 min in this case (the period of the semidiurnal lunar tide). However, the lunar orbit does not lie in the plane of the equator and the Moon is alternately to the north and to the south of the equator. The daily rotation of the Earth about an axis inclined to the lunar orbital plane introduces an asymmetry in the tides. This asymmetry is apparent as an inequality of the two successive cycles within 24 h 50 min.

Similarly, the Sun causes a semidiurnal solar tide with a 12 h period and a diurnal solar tide with a 24 h period. In a complete description of the local variations of the tidal forces, other partial tides still play a role because of further inequalities in the orbital motions of the Moon and the Earth. In particular, the elliptical shape of the Moon's orbit produces a 40 per cent difference between the lunar tidal forces at the perigee and apogee of the orbit. Also the inclination of the Moon's orbit varies periodically in the interval 18.3°–28.6°, causing a partial tide with a period of 18.6 years. The interference of the Sun-induced tidal forces with the Moon-induced tidal forces (the lunar forces are about 2.2 times as strong) causes the regular variation of the tidal range between spring tide, when the range has its maximum (occurring at a new Moon and at a full Moon, when the Sun and Moon are in the same or in the opposite directions), and neap tide, when the range has its minimum (which occurs at intermediate phases of the Moon). The amplitude of a spring tide may be 2.7 times the amplitude of a neap tide.

Because the Earth is not surrounded by an uninterrupted water envelope of equal depth, but rather has a very irregular geographic alternation of land and seas with complex floor geometry, the actual response of the oceans and seas to the tidal forces is extremely complex. In enclosures formed by gulfs and bays, the local tide is generated by an interaction with the tides of the adjacent open ocean. Such a tide often takes the form of a running tidal wave that circulates within the confines of the enclosure. In some nearly enclosed seas, such as the Mediterranean, Black and Baltic seas, a steady-state oscillation in the form of a standing wave, or tidal seiche, may be generated by the tidal forces. In these seas, the tidal range of sea level is only on the order of centimeters. In the open ocean, it generally is on the order of decimeters.

In bays and adjacent seas, however, the tidal range may be much greater because the shape of a bay or adjacent sea may favor the enhancement of the tide inside. In particular, there may be a resonance response of the basin concerned with the tide. Tides are most easily observed along seacoasts, where the amplitudes are exaggerated. When tidal currents run into the shallow waters of the continental shelf, their rate of advance is reduced, the energy accumulates in a smaller volume and the rise and fall are amplified. The details of tidal motions in coastal waters, particularly in channels, gulfs and estuaries, depend on the details of coastal geometry and water-depth variation over a complex sea floor. Tidal amplitudes and phase lags, the contrast between spring and neap tides, and the variation of the times of high and low tide all change widely from place to place.

For the aforementioned reasons, a purely theoretical calculation of the times and heights of tides at a particular location is practically impossible. Nevertheless, for a given place on a coast, the tides can be quite successfully predicted on the basis of accumulated long-term observations of the tides at the place concerned. The analysis of the observations relies on the fact that any tidal pattern in time is a superposition of variations associated with periodicities in the motions of the Moon and the Sun relative to the Earth. The periods involved are the same everywhere on the Earth, but the relative amplitudes and phases of their contributions are highly variable from one place to another. Observations over a sufficient time make it possible to calculate which contributions are significant at a particular location and, thus, to forecast tidal times and heights. It is common that 40 harmonic components may be significant for practical calculations at one location.

### 11.13.11 The evolution of orbital motions and spins of celestial bodies induced by tidal forces

Besides perturbing the satellites' orbits, the gravitational field of one body in an orbit about another tidally distorts the shape of the other body. The solid body of the Earth also experiences twice-daily tides with a maximum amplitude of about 30 centimeters.

When the forced motion occurs exactly in the same or opposite phase with respect to the driving force, no energy exchange occurs on average between the external source and the oscillatory system. However, it is well known from astronomical observations that the Moon is slowly receding from the Earth, with an accompanying increase in its orbital period (the month). While the Moon's orbital momentum increases, the Earth's spin momentum decreases: the Earth's daily rotation slows down, so that the day becomes longer. To explain the secular variation (the retardation) of the Earth's axial rotation under the tidal forces, we have to take friction into account.

One may wonder why the dissipation of mechanical energy in the tides has a scale that seems very modest. The point is that only the wave circulates around the globe,

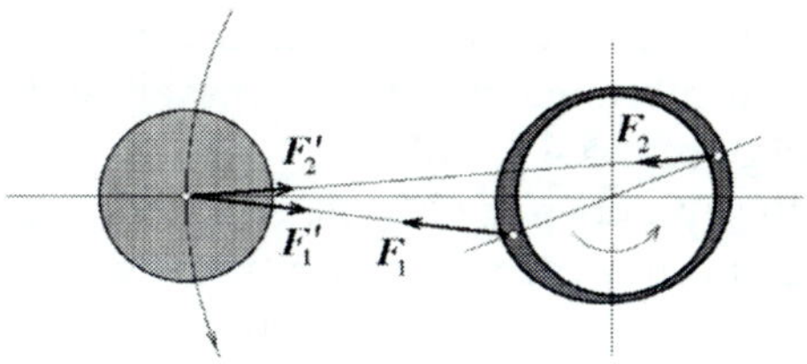


**Figure 11.20.** The gravitational interaction between the Moon and the tidal bulges.

not the water itself. The phase lag $\delta$ of the steady-state forced oscillation behind the periodic driving force is determined by (11.107). For the mode of oscillations in which we are interested, this phase-frequency characteristic is almost a step function (zero for $\omega < \omega_0$, that is, for $T > T_0$, and $-\pi$ otherwise). Only near resonance ($\omega \approx \omega_0$) is this step slightly smoothed over. Therefore the displacement of the tidal water bulges from the line perpendicular to the Sun–Earth (Moon–Earth) axis is very small.

However, this displacement, which destroys the symmetry of the system (figure 11.20), is absolutely necessary in principle in order that the driving tidal forces be capable of maintaining the circulating tidal wave (that is, of preventing it from damping out). If the Earth is taken as the reference frame, we can see that by virtue of this phase shift and the corresponding displacement of bulges, the tidal forces exert a retarding torque relative to the Earth's axis and thus perform non-zero net work on the system. This work compensates for the frictional losses experienced by the tidal traveling surface wave and measures the gradual reduction of the mechanical energy of the system. The energy is provided by the axial rotation (spin) of the Earth. Hence the spin secularly slows down and the angular momentum of the axial rotation diminishes.

Looking at the whole system from the inertial reference frame, we should remember that the Sun (Moon) interacts with the Earth only by its central gravitational force. If the bulges were oriented exactly along or perpendicularly to the Sun–Earth (Moon–Earth) axis, this gravitational force would not exert a torque on the Earth. If we consider the gravitational forces $\mathbf{F}_1$ and $\mathbf{F}_2$ (figure 11.20) exerted on the bulges, we conclude that the retarding torque about the Earth's axis, which slows down the axial rotation, is due to the above mentioned displacement of the bulges which destroys the symmetry of the system with respect to the Earth–Sun (Earth–Moon) line.

However, the total torque of the central gravitational field of the Sun (Moon) exerted on the Earth and the bulges of its liquid shell, measured relative to the Sun (or to the Moon for Moon-induced tides), is zero. Hence the total angular momentum of the system is conserved, as it should be in any closed system. The diminishing of the Earth's spin due to tidal friction means that the orbital momentum of the system slowly increases during tidal evolution. The Earth's orbit gradually expands. The lack of symmetry (produced by tidal friction) does not influence the conservation of total angular momentum, although it causes a slow secular redistribution of the angular momentum between the spin and the orbital motion. As the orbit expands, the mechanical energy of the orbital motion also increases. This additional mechanical energy, as well as the dissipated energy, is borrowed from the energy of axial rotation.

This conclusion about expanding the Moon's orbit, derived from the conservation of angular momentum, is often encountered in the literature. Although quite convincing, it nevertheless leaves the actual mechanism unexplained. To understand the physical reason for this phenomenon, it helps to take the forces into account. If we consider the properties of the gravitational forces $\mathbf{F}_1$ and $\mathbf{F}_2$ (see figure 11.20) exerted on the Moon by the Earth's tidal bulges and their influence on the orbital motion, we draw attention to a subtle peculiarity that deserves discussion. While the orbit expands, the orbital velocity of the Moon diminishes. However, from the asymmetry in the configuration that is responsible for the evolution, we can conclude that the resultant gravitational force exerted on the Moon by the tidal bulges (the sum of $\mathbf{F}_1$ and $\mathbf{F}_2$) has a component which is directed forward, in the direction of the orbital motion. How can this accelerating force slow down the orbital motion? All authors who write about tidal evolution leave this question unanswered.

This situation is similar to the widely known paradox of an Earth-orbiting satellite in a circular orbit that gradually descends in the rarified upper atmosphere. Intuitively we expect that the weak atmospheric drag should slow down the satellite, but instead, the satellite gains speed as its orbit gradually decreases. Because of air resistance, the satellite is accelerated in the direction of its motion, as if the retarding force of air resistance were pushing the satellite forward. A physical explanation of this so-called aerodynamical paradox of the satellite can be found in section 4.2.2.

To understand the slowing down of the Moon during tidal evolution, we must take into account that the Moon gradually spirals away from the Earth and its orbit spreads out, so that the actual motion of the Moon occurs along an expanding spiral. A portion of this trajectory (with a strongly exaggerated expansion) is shown schematically in figure 11.21. Because of this expansion, the perpendicular to the trajectory is directed not to the center of the Earth but rather slightly in front of the center. Therefore the main gravitational pull $\mathbf{F}$ exerted on the Moon by the Earth has a retarding tangential component $\mathbf{F}_\tau$ directed back along the trajectory. This component is greater in magnitude than the forward-directed tangential component of $\mathbf{F}_1$ and $\mathbf{F}_2$ (see figure 11.20) exerted on the Moon by the tidal bulges (this

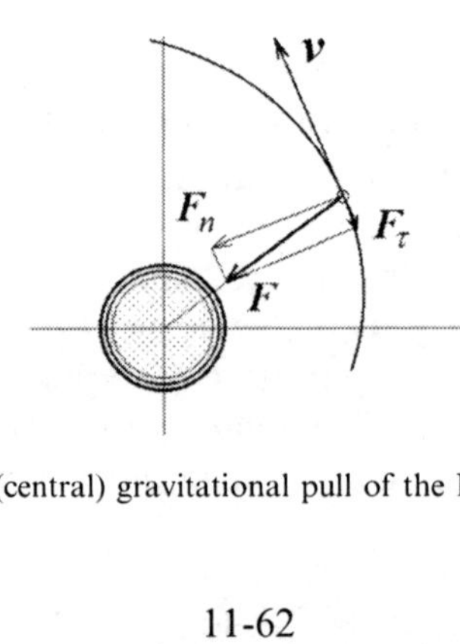


**Figure 11.21.** The main (central) gravitational pull of the Earth exerted on the Moon.

component is not shown in figure 11.21). Hence the total tangential acceleration of the Moon is directed against the velocity.

Generally, in order to explain tidal evolution, that is, the reduction of spin and the secular variation of the orbits of gravitationally coupled celestial bodies, it is necessary to take into account both the dynamic distortion of the spherical shape of the body (and of its liquid shell, if any) under the tidal forces and the additional displacement of the bulges caused by tidal friction. The non-uniform gravitational field of one body in an orbit about another distorts the shape of the second. The dissipation of energy stored in the resultant tidal distortions leads to a coupling that causes secular changes (always in the same direction) in the orbit and in the spins of both bodies. Retardation of the axial rotation and evolution of the orbit will continue until the axial rotation is synchronous with the mean orbital revolution.

This effect is vital for an understanding of the history of the Earth and Moon. That the Moon always keeps the same face turned toward the Earth is attributed to the past effects of tidal friction in the Moon. The dissipation of tidal energy on the Earth results in a slowing of the Earth's axial rotation while the Moon's orbit is gradually expanding. Both the currently observed increase in the length of the day of 0.0016 second per century and the recession of the Moon of 3 to 4 cm per year are understood as consequences of the tides raised by the Moon on the Earth. The Moon will continue to recede and slow down, and the angular momentum will be exchanged, until the Earth presents the same face toward the Moon. Billions of years from now the Moon will be so far from the Earth that the duration of the month will be equal to the duration of the day. The tidal evolution of the system ends with synchronization of the axial rotation of both orbiting bodies with their orbital revolution. The length of both the day and month in this final state of coherent rotation will be approximately 50 present days, as can be calculated on the basis of angular momentum conservation. Similarly, tidal effects on the Earth influence its axial rotation and its orbital revolution around the Sun.

The variation of the orbital motion during the tidal evolution is also interesting in the sense that it disproves a common belief that internal forces cannot influence the motion of the center of mass of a system. This statement is true only for a system moving in a homogeneous external field. The aforementioned expansion of the Earth's orbit during tidal evolution is actually caused (though indirectly) by internal forces, namely by gravitational forces between the ocean water and the hard body of the Earth (Earth's self-gravity), and frictional forces. The internal forces change the configuration of the system (consisting of the Earth and the ocean water) that moves in the non-homogeneous external gravitational field. Therefore these internal forces change (indirectly) the resultant external gravitational force exerted on the system. These variations of the external gravitational force modify the orbital motion of the system. During tidal evolution, the internal forces of mutual gravitation together with friction cause the redistribution of masses (of the Earth and tidal bulges) that move in the external non-homogeneous central field.

Tidal dissipation accounts for the current states of axial rotation of several planets, the spin states of most of the planetary satellites and the spins and orbits of close binary stars. For example, all the major and close planetary satellites in the

Solar System (with the exception of Saturn's satellite Hyperon) are observed to be rotating synchronously with their orbital motion. The distant planet Pluto and its satellite Charon are the pair in the Solar System that has almost certainly reached the end point where further tidal evolution has ceased. In this state the orbit is circular, with both bodies rotating synchronously with the orbital motion and both spin axes perpendicular to the orbital plane. Similarly, many close binary stars are observed to have circular orbits and synchronized spins, providing numerous examples of evolution under tidal forces elsewhere in the Milky Way. The role of tides in the cosmogony was first recognized by the astronomer George Darwin (1845–1912) who developed a theory of heavenly evolution under tidal friction.

Another interesting manifestation of the tidal forces is the *Roche limit*, the minimum distance to which a large (natural) satellite can approach its primary body without being torn apart by tidal forces. To calculate this critical distance $R$, we equate the vertical tidal force, (11.88) with $\theta = 0$ or $\theta = \pi$, exerted on a mass point on the surface of a satellite of radius $r_{sat}$ and mass $m_{sat}$ by its primary of mass $M$, and the force of self-gravitation of the satellite (i.e. the force of gravitational attraction of this mass point $m$ to the satellite):

$$2\,\frac{GmM}{R^3}\,r_{sat} = \frac{Gmm_{sat}}{r_{sat}^2}, \qquad (11.109)$$

whence

$$R = r_{sat}\sqrt[3]{\frac{2\,M}{m_{sat}}} = r_{pl}\sqrt[3]{\frac{2\rho}{\rho_{sat}}}. \qquad (11.110)$$

In the latter expression for the critical distance, $r_{pl}$ is the radius of the primary, $\rho$ is its mean density and $\rho_{sat}$ is the satellite's mean density. If the satellite and its primary are of similar composition ($\rho \approx \rho_{sat}$), the theoretical limit is about $\sqrt[3]{2} = 1.26$ times the radius of the larger body.

The limit was first calculated by the French astronomer Edouard Roche (1820–83). The above calculation of the Roche limit assumes that the satellite remains spherical: it does not take into account the inevitable distortion of the satellite's shape in the presence of strong tidal forces. A more accurate calculation, which treats the satellite as a self-gravitating incompressible fluid, yields about twice as large a value for the critical distance $R$.

We note that small orbital bodies such as rocks, or even small moons, can survive intact within the Roche radius because they are held together primarily by internal tensile forces rather than by gravitational attraction. However, this mechanism becomes progressively less effective as the size of the body in question increases. Not surprisingly, all large planetary moons occurring in the Solar System have orbital radii which exceed the relevant Roche radius, whereas all planetary ring systems (which consist of myriads of small orbiting rocks) have radii which lie inside the Roche radius. The famous rings of Saturn lie inside Saturn's Roche limit and may be the debris of a demolished moon. Artificial satellites are too small to experience substantial tidal stresses.

# Glossary

- **Acceleration of free-fall.** The *acceleration of free-fall* is the acceleration experienced by any body moving freely under the action of only the gravitational field. The acceleration of free-fall is also called the *acceleration due to gravity*. In the same gravitational field all bodies, independent of their masses, fall with the same acceleration. Moreover, the acceleration is independent of the nature of the falling body. This property has been established experimentally with great precision. The property indicates that for all bodies the *inertial mass* $m_{in}$ (the mass that appears in Newton's second law of motion) is strictly proportional to the *gravitational mass* $m_{gr}$ that characterizes the force experienced by the body in the gravitational field (the mass that appears in Newton's law of gravitation). Although the two masses characterize different properties of the body, they are physically identical (equivalent) and thus can be measured in the same units ($m_{gr} = m_{in}$). Einstein came to the conclusion that this exact coincidence of the two masses cannot be accidental, but rather indicates the equivalence of inertial and gravitational phenomena.
- **Aphelion.** The *aphelion* is the point in the finite orbit of a planet, comet, asteroid, or artificial satellite in solar orbit at which the orbiting body is farthest from the Sun. At the *perihelion* the orbiting body is nearest to the Sun.
- **Apogee.** The *apogee* is the point in the orbit of the Moon, or of an artificial Earth satellite, at which it is furthest from the Earth. At the *perigee* of an orbit, the orbiting body is nearest to the Earth.
- **Asteroid.** *Asteroids* (or *minor planets*) are small bodies that revolve around the Sun between the orbits of Mars and Jupiter in a region called the *asteroid belt*. The size of asteroids varies from about 1000 km for the largest (Ceres) to less than 1 km in diameter. It is estimated that there are about 500 asteroids with diameters in excess of 100 km.
- **Chaos.** *Chaos* is an irregular, seemingly random behavior occurring in a system governed by deterministic laws. Planetary dynamics is one of numerous examples of chaos in physics. To exhibit chaos, the system must be described by *non-linear* differential equations involving several variables. Such systems may be very sensitive to the initial conditions, so that a very small initial difference may make an enormous change in the future state of the system.
- **Circular velocity.** *Circular velocity* $v_c$ is the speed that a celestial body or a satellite must have in order to stay in a circular orbit of a definite radius (around the Sun or some other celestial body). Its value can be calculated on the basis of Newton's second law by equating the centripetal acceleration of the body (multiplied by the mass of the body) to the gravitational force of attraction exerted on the orbiting body by the central body:

$$v_c = \sqrt{\frac{GM}{r}}, \qquad v_c = \sqrt{\frac{gR^2}{r}}.$$

 

Here $G$ is the gravitational constant, $M$ is the mass of the central body and $r$ is the radius of the circular orbit. The second expression gives the circular velocity for an Earth-orbiting satellite. In it, $g = 9.8\ \mathrm{m\ s^{-2}}$ is the *acceleration of free-fall* near the Earth's surface, $R = 6370$ km is the Earth's radius and $r$ is the radius of the circular orbit.

For a given central body, the circular velocity is inversely proportional to the square root of the radius $r$ of the orbit. The circular velocity is independent of the mass $m$ of the orbiting body provided $m \ll M$. Otherwise we should consider the orbital motion of both bodies around the center of mass of the system rather then the motion of one of them around the other.

- **Comet.** A *comet* is a small celestial body that travels around the Sun either in a highly eccentric elliptical orbit, or in a parabolic (or hyperbolic) trajectory. *Short-period comets* have orbital periods of less than 150 years. The periods of *long-period* comets may exceed 100000 years. The most famous short-period comet is *Halley's comet*, whose period is about 76 years. Its last visit was in 1986. The comet is named after Edmund Halley (1656–1742) who first calculated its orbit.

  Typical comets have a *nucleus* of ice and dust, a nebulous cloud of gas and dust (called the *coma*) that surrounds the nucleus, and the comet *tail*, also of gas and dust, which only appears when the comet is near the Sun.
- **Conic sections.** A *conic* or *conic section* is a figure formed by the intersection of a plane and a circular cone. If the intersecting plane is perpendicular to the axis of the cone, the figure formed is a circle. If the intersecting plane is inclined to the axis at an angle greater than half the apex angle of the cone, it is an ellipse. If the plane is parallel to the sloping side of the cone, the figure is a parabola. If the plane cuts both halves of the cone, two branches of a hyperbola are formed by the intersection.
- **Ellipse.** An *ellipse* is a *conic* (or *conic section*) that can be defined as a plane curve whose points satisfy the following property: the sum of the distances from any point of the curve to the two fixed points (called the *foci* of the ellipse) is constant.

  The equation of an ellipse in polar coordinates can be written as:

$$r(\varphi) = \frac{p}{1 + e\cos(\varphi - \varphi_0)}.$$

The quantity $p$ is called the *semilatus rectum* or the *orbital parameter* (or the *focal parameter*) of the conic and the dimensionless parameter $e$ is called the *eccentricity* of the conic section. For $e = 0$ both foci of the ellipse coincide and the curve becomes a circle. For $e = 1$ the ellipse degenerates into a straight line joining the foci.

The ellipse has two axes of symmetry (the *major* and *minor* axes), which are mutually perpendicular. The area bounded by an ellipse equals $\pi ab$, where $a$ and $b$ are the *semimajor* and *semiminor* axes. The foci of an ellipse are at a distance $ea$ from its center, where $e$ is the eccentricity.

- **Escape velocity.** The *escape velocity* is the minimum speed needed by a space vehicle to escape from the gravitational field of the Earth, Moon or other celestial body. The gravitational potential energy of a body with mass $m$ in the central gravitational field created by a spherically symetric celestial body of mass $M$ equals $-GmM/r$. To escape from the gravitational field, the body must have a kinetic energy that exceeds the magnitude $|GmM/r|$ of this potential energy. Hence the escape velocity is inversely proportional to the square root of the distance $r$ from the celestial body.
- **Hodograph of the velocity vector.** Generally, the *hodograph of a vector* is the curve generated by the end of the changing vector while the origin of the vector remains at the same point. For example, the spatial trajectory of a moving particle is the hodograph of its radius vector.

  The hodograph of the velocity vector of a body that moves in a central Newtonian gravitational field is a circle (or part of a circle). This property of an arbitrary Keplerian motion is illustrated by the simulation program 'Hodograph of the Velocity'.
- **Hyperbola.** A *hyperbola* is a *conic section* that can be defined as a curve with the following property: the difference in the distances of any point of a hyperbola from two fixed points (called the *foci* of the hyperbola) is constant. The equation of a hyperbola in polar coordinates is:

  $$r(\varphi) = \frac{p}{1 + e \cos \varphi},$$

  where the *eccentricity* $e > 1$. (Compare this expression with that for the equation of an ellipse, where $e < 1$.) Keplerian motion with a positive value of the total energy occurs along a hyperbola.
- **Keplerian motion.** The motion under a sole central force of attraction that decreases as the square of the distance from the center of force. Examples of Keplerian motion are the unperturbed motion of a planet around the Sun and the motion of satellites around the Earth (and other planets). Keplerian motion obeys *Kepler's laws* and arises from the *Newtonian force of gravitation.*
- **Kepler's laws.** Three laws that govern the motion of planets around the Sun:
  1. The orbits of the planets are ellipses with the Sun at one focus of the ellipse (*Kepler's first law*).
  2. Each planet revolves around the Sun so that the radius vector connecting the Sun to the planet sweeps out equal areas in equal times (*Kepler's second law*).
  3. The ratio of the square of each planet's period to the cube of its distance from the Sun is a constant for all the planets (*Kepler's third law*).

These laws of planetary motion were established by Johannes Kepler (1571–1630) in about 1610 on the basis of astronomical observations made by Tycho Brahe (1546–1601). The laws hold for any *Keplerian motion* governed by a central force that decreases as the square of the distance from the center of force.

- **Newton's law of gravitation.** For any two point masses $m_1$ and $m_2$, the force of gravitational attraction decreases as the square of distance $r$ between the points:

$$F_r(r) = -G\frac{m_1 m_2}{r^2}, \qquad \mathbf{F}_{12} = -G\frac{m_1 m_2}{r^2}\frac{\mathbf{r}_{12}}{r}.$$

  Here $G$ is the gravitational constant and $r$ is the distance between the interacting bodies (mass points), whose masses are $m_1$ and $m_2$. The first expression gives the projection of the gravitational force onto the radius vector $\mathbf{r}$ (the negative sign shows that it is the force of attraction directed toward the origin). In the second expression (which gives the vector $\mathbf{F}_{12}$ of the gravitational force exerted on mass $m_1$ by $m_2$), the vector $\mathbf{r}_{12}$ is directed from $m_1$ to $m_2$.

  This expression is equally valid for the gravitational force exerted on a small body (point mass) by an arbitrarily large body with spherically symmetric mass distribution and for the force of gravitational interaction between two spherically symmetric massive bodies, such as stars or planets. In this case $r$ is the center-to-center distance between the bodies.
- **Osculating orbit.** In the stationary central gravitational field whose force is decreasing according to the inverse-square law, an orbit of a satellite (or a planet) is exactly an ellipse. When small perturbations are taken into account, it is convenient to approximately consider the orbit as an ellipse whose parameters are defined by the instantaneous values of the position and velocity vectors. Small perturbations cause the parameters of the ellipse to vary slowly. Such an ellipse with varying parameters is called an *osculating orbit*. That is, the osculating ellipse is that elliptical orbit that would be assumed by the body if all the perturbing forces were suddenly turned off.
- **Parabola.** A *parabola* is a *conic section* that can be defined as a curve with the following property: the distance of any point of a parabola from the *focus* is equal to its perpendicular distance from a straight line called the *directrix* of the parabola. Keplerian motion with a total energy of zero occurs along a parabola.
- **Perigee.** See *Apogee*.
- **Perihelion.** See *Aphelion*.
- **Perturbation.** A *perturbation* is a departure of a celestial body from the trajectory or orbit it would follow were it to move only under the influence of a single central gravitational force that decreases as the square of the distance from the center of force. Such a force is described by *Newton's law of gravitation*. According to *Kepler's laws*, a single planet orbiting the Sun or a satellite orbiting the Earth moves in an elliptical orbit. But in fact, planets are perturbed from elliptical orbits by the gravitational forces exerted on them by other planets. Similarly, satellites orbiting the Earth are perturbed by the gravitational effect of the Sun and the Moon (by the tidal forces), and by the deviation of the Earth's shape from a sphere. Parabolic or hyperbolic trajectories of comets are perturbed when they pass close to planets.

- **Sidereal period.** The *sidereal period* is the time taken for a planet (or satellite) to complete one revolution of its orbit measured with reference to the background of the stars.
- **Sphere of gravitational action.** A concept used in the approximate treatment of the restricted three-body problem by the method of joined conic sections. For example, the (passive) motion of a spacecraft en route to another planet within the sphere of gravitational action of the Earth can be treated as a geocentric Keplerian motion almost unperturbed by the Sun. After the spacecraft leaves the sphere, its motion is approximately a heliocentric Keplerian motion. After the spacecraft enters the sphere of gravitational action of the target planet, its motion is treated as a planetocentric Keplerian motion. On the boundary of the sphere of gravitational action, the two Keplerian orbits are joined by the transformation of the coordinates and velocity of the spacecraft from one frame of reference to the other, and these new values are regarded as the initial conditions for the continued Keplerian motion in the new frame of reference.

  The boundary of the sphere of gravitational action of a planet relative to the Sun is assumed by convention to be the surface on which the ratio of the perturbational acceleration to the main acceleration for the planetocentric motion equals such a ratio for the heliocentric motion. The value

  $$r = R\left(\frac{m}{M}\right)^{2/5}$$

  gives the radius of the sphere of gravitational action of a small body (a planet of mass $m$) relative to the massive body (the Sun, mass $M$). Here $R$ is the distance between the bodies.
- **Synodic period.** The *synodic period* is the mean time that elapses between two successive identical configurations of a planet and the Sun as seen from the Earth, e.g. the mean time $S$ between successive returns to the opposition. For inferior planets $1/S = 1/T - 1/E$, where $T$ is the *sidereal period* of the planet and $E$ is the sidereal period of the Earth. For superior planets $1/S = 1/E - 1/T$.

Lightning Source UK Ltd.
Milton Keynes UK
UKOW03n0919090115

244206UK00001B/13/P

9 780750 311014